172 Topics in Current Chemistry

Cyclophanes

Editor: E. Weber

With contributions by
H. Hart, S. Inokuma, H.-B. Mekelburger,
J. Nishimura, S. Sakai, A. Schröder,
J. Schulz, Y. Tobe, T. K. Vinod, F. Vögtle

With 98 Figures and 21 Tables

Springer-Verlag
Berlin Heidelberg GmbH

This series presents critical reviews of the present position and future trends in modern chemical research. It is addressed to all research and industrial chemists who wish to keep abreast of advances in their subject.

As a rule, contributions are specially commissioned. The editors and publishers will, however, always be pleased to receive suggestions and supplementary information. Papers are accepted for "Topics in Current Chemistry" in English.

ISBN 978-3-662-14911-9   ISBN 978-3-540-48587-2 (eBook)
DOI 10.1007/978-3-540-48587-2

Library of Congress Catalog Card Number 74-644622

© Springer-Verlag Berlin Heidelberg 1994

Originally published by Springer-Verlag Berlin Heidelberg New York in 1994
Softcover reprint of the hardcover 1st edition 1994

The use of general descriptive names, registered names, trademarks, etc. in this publication does not imply, even in the absence of a specific statement, that such names are exempt from the relevant protective laws and regulations and therefore free for general use.

Typesetting: Macmillan India Ltd., Bangalore-25
Offsetprinting: Saladruck, Berlin; Bookbinding: Lüderitz & Bauer, Berlin
SPIN: 10474057   51/3020 - 5 4 3 2 1 0 - Printed on acid-free paper

Guest Editor

Prof. Dr. *E. Weber*
Institut für Organische Chemie, Bergakademie Freiberg
Technische Universität, Leipziger Str. 29, 09596 Freiberg

Editorial Board

Attention
all "Topics in Current Chemistry" readers:

A file with the complete volume indexes Vols.22 (1972) through 171 (1994) in delimited ASCII format is available for downloading at no charge from the Springer EARN mailbox. Delimited ASCII format can be imported into most databanks.

The file has been compressed using the popular shareware program "PKZIP" (Trademark of PKware Inc., PKZIP is available from most BBS and shareware distributors).

This file is distributed without any expressed or implied warranty.

To receive this file send an e-mail message to:
SVSERV@VAX.NTP. SPRINGER.DE
The message must be:"GET/CHEMISTRY/TCC_CONT.ZIP".

SVSERV is an automatic data distribution system. It responds to your message. The following commands are available:

HELP	returns a detailed instruction set for the use of SVSERV
DIR (name)	returns a list of files available in the directory "name",
INDEX (name)	same as "DIR",
CD <name>	changes to directory "name",
SEND <filename>	invokes a message with the file "filename",
GET <filename>	same as "SEND".

For more information send a message to:
INTERNET:STUMPE@SPINT. COMPUSERVE.COM

Preface

In 1957, *D. J. Cram* and *H. Steinberg* [(1957) J Am Chem Soc 73: 569] reported on the synthesis of a compound in which two benzene rings are held face-to-face by two dimethylene bridges between positions 1 and 4 of the aromatic nuclei, respectively. Attractively naming this compound as [2.2] paracyclophane led to the development of the cyclophane nomenclature by *F. Vögtle* and *P. Neumann* [(1970) Tetrahedron 26: 5847]. According to this systematization, all molecules with at least one *aromatic* ring bridged by at least one *aliphatic* n-membered bridge may be called *Cyclophanes* which is a word contraction of *cycloph*enylene alk*an*e.

Over the years, this particular class of compounds has mushroomed to give an enormous variety of compounds all having the characteristic feature of a cyclophane. Not only do cyclophanes enjoy great popularity in general, but they have also invaded many areas of chemistry. There is almost no chemical branch where cyclophane structures are not met, mostly at a central point whether in a scientific or an application sense.

Perhaps the most efficient stimulus in cyclophane chemistry goes back to the discovery of crown ethers by *C. J. Pedersen* in 1967 [(1967) J Am Chem Soc 89: 7017] being the starting signal for a very promising field of research called *Host-Guest* or *Supramolecular Chemistry*. Actually the first crown ether that was synthesized, dibenzo-18-crown-6, was a cyclophane.

Today, Supramolecular Chemistry would be unthinkable without the structural building blocks of cyclophanes. Nevertheless, traditional subjects attached to cyclophanes such as molecular strain, transannular interactions and other things are still of current interest.

The present book provides a state-of-the-art view on some of these topics including topological, conformational and host-guest problems written by five recognized experts. Because of this, the book lays no claim to being a quantitative overview of cyclophane chemistry. However, in this latter respect, the reader is referred to the excellent monographs produced by *F. Diederich* [(1991)

Cyclophanes. Royal Society of Chemistry, Cambridge] and *F. Vögtle* [(1993) Cyclophane Chemistry. Wiley, Chichester]. They may serve as encyclopedia of cyclophane chemistry prior to 1992. In addition, two volumes on 'Cyclophanes' have already appeared in the present series [(1983 Top Curr Chem 113 and 115].

To conclude these words of introduction, I wish to thank all contributors for their help and participation in writing this book.

Freiberg, July 1994   Edwin Weber

Table of Contents

Strained [n]Cyclophanes

Yoshito Tobe

Department of Chemistry, Faculty of Engineering Science, Osaka University,
1-1 Machikaneyama, Toyonaka, Osaka 560, Japan

Table of Contents

Strained [n]paracyclophanes and [n]metacyclophanes with a short bridge serve as good basis for the relationship between aromaticity and planarity of the π system. They have also been receiving challenges as synthetic targets because of their aesthetically beautiful architecture. Remarkable advances have been achieved in this field during the last decade; the limit of their existence has been explored extensively owing to the development of new synthetic methodologies. The smallest representatives that are stable at room temperature, [6]paracyclophane and [5]metacyclophane, as well as their many derivatives were prepared and their geometries, spectral properties, and reactivities have been investigated in detail. The lower homologues, [5]paracyclophane, [4]paracyclophane, and [4]metacyclophane, have also been characterized chemically or spectroscopically. Considerable advances have also been achieved in the [n]cyclophanes of condensed benzenoid aromatics which exhibit extraordinary reactivities. This chapter deals with the recent advance in the synthetic methodologies, theoretical studies, and studies on geometries and reactivities of strained [n]cyclophanes with the main focus on those with the number of the bridge atoms (n) less than six in the para series and less than five in the meta series.

List of Abbreviations

DCNA	Dicyanoacetylene
DMAD	Dimethyl acetylenedicarboxylate
DMSO	Dimethyl sulfoxide
FVP	Flash vacuum pyrolysis
TAD	N-Phenyl-1, 2, 4-triazoline-3, 5-dione
TCNE	Tetracyanoethylene
TFA	Trifluoroacetic acid
TMEDA	Tetramethylethylenediamine
TMSCl	Trimethylsilyl chloride

1 Introduction

Chemists naturally ask themselves "How much bending can a benzene ring withstand without giving up its aromatic character? [1]" or, more simply, "How bent can a benzene ring be? [2]" Strained [n]paracyclophanes and [n] metacyclophanes with a short bridge can serve as good basis for these questions and not only from this theoretical interest, strained [n]cyclophanes have also been a challenge as synthetic targets, because they are aesthetically beautiful, the simplest of cyclophanes. In this respect, the chemistry of strained [n]cyclophanes continues to represent one of the major topics in the chemistry of cyclophanes [3]. In particular, remarkable advances have been achieved in this field during the last decade; the limit of their existence has been explored extensively owing to the development of new synthetic methodologies. [6] Paracyclophane (1a) and [5]metacyclophane (4a) (Structures 1) are the smallest representatives in each series that are stable at room temperature. Adequate quantities of the parent hydrocarbons as well as their many derivatives have been prepared and their geometries, spectral properties, and reactivities have been investigated in detail. The lower homologues of the [n]paracyclophane series, [5]paracyclophane (2a) and [4]paracyclophane (3a) (Structures 1), were characterized as reactive intermediates by both spectroscopic and chemical methods. [4]Metacyclophane (**5**) was also postulated as being a reactive intermediate, however, it has not been characterized spectroscopically yet.

1a n=6   **4a** n=5
2a n=5   **5** n=4
3a n=4

Structures 1

As a natural extension of the chemistry of [n]cyclophanes, remarkable advances have been achieved in the chemistry of [n]cyclophanes of condensed benzenoid aromatics. Because of the diversity of the π bond orders and large HOMO/LUMO electron densities, condensed aromatics are more sensitive to strain imposed by the short bridge than the corresponding benzologue. From this point of view, the naphthalene and anthracene analogues of [6]paracyclophane were synthesized and their geometries and reactivities were investigated. Particularly noteworthy is that they exhibit extraordinary reactivities.

This contribution deals with the recent advances in the synthetic methodologies, geometrical and theoretical studies, and studies on reactivities of strained [n]cyclophanes. It is generally accepted that "strained" [n]cyclophanes include those with the number of the bridge atoms (n) less than eight in the *para* series and less than seven in the *meta* series, since their spectroscopic properties and chemical reactivities start to deviate from those of planar aromatic compounds [3e, f]. However, because of space, I will focus on the smallest representatives, [6]paracyclophane (**1a**) and [5]metacyclophane (**4a**), their lower homologues, and their analogues in condensed benzenoid aromatics, and will refer to the larger [n]cyclophanes only when comparison is appropriate.

2 Synthesis of Strained [n]Cyclophanes

2.1 Synthesis of Strained [n]Paracyclophanes

The substantial amount of strain in [6]paracyclophane (**1a**) (Structures 1) and its lower homologues precluded ordinary synthetic methods for cyclophane synthesis. Viable alternative should therefore be based on transformation from high energy species such as strained molecules or reactive intermediates like carbenes and molecules in an electronically excited state. Four methods have been developed so far for the synthesis of **1a**, all fulfill this criterion.

Jones developed a new fascinating method for [n]paracyclophane synthesis based on spiro carbene rearrangement. The first synthesis of **1a** was achieved two decades ago by this method through rearrangement of the carbene species generated by thermolysis of the lithium salt of the spiro dienone tosylhydrazone **6** (Scheme 1) [1]. Although this work was epoch-making as the first synthesis of **1a**, the formation of by-products hampered further study of **1a**.

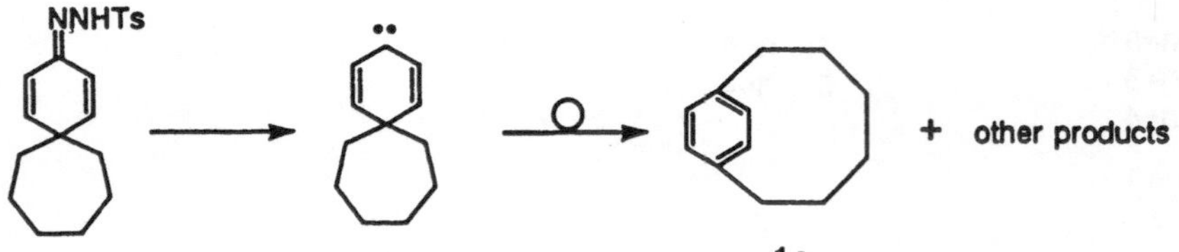

Scheme 1.

Soon after the first synthesis, a notable landmark, the second route to **1a**, was achieved by Jones and Bickelhaupt; they reported that the thermal valence isomerization of Dewar benzene type isomer of **1a**, [6.2.2]propelladiene (**8a**), proceeded smoothly, giving **1a** quantitatively (Scheme 2) [4]. This reaction provided an entirely new general synthetic method for [n]paracyclophanes. However, the overall sequence of reactions was not enough to give a large quantity of **1a** at that time, because the precursor **8a** was prepared as a minor product of silver-(I)-catalyzed isomerization of the bicyclopropenyl derivative 7.

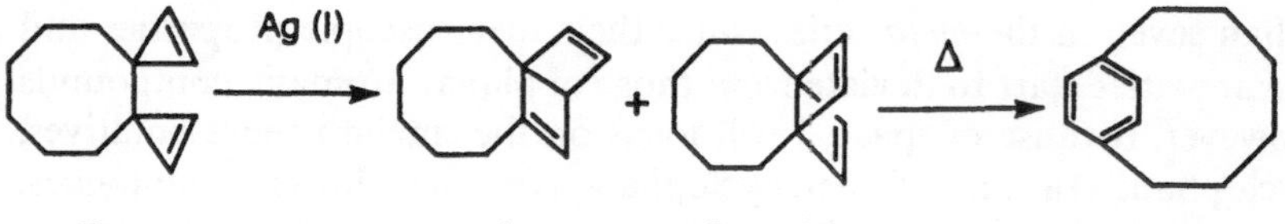

Scheme 2.

We took advantage of this valence isomerization method; we prepared the precursor **8b** efficiently by using enone-olefin photocycloaddition and subsequent ring contraction. Thermal isomerization of **8b** proceeded quantitatively, giving methoxycarbonyl-substituted cyclophane **1b** (Scheme 3) [5]. Bis(methoxycarbonyl) derivative **1c** [6] and diketo derivative **1e** (Structures 2) [7] were also prepared by a similar procedure. Moreover, [6]paracycloph-3-ene (**9a**) and its derivatives **9b** and **9c** (Structures 2) were synthesized by using this method [8]. Since the bridge of this system is shorter than that of **1a**, **9a** represents the most strained [n]paracyclophane that has been isolated. Gleiter also prepared the tetrasubstituted [6]paracyclophane derivative **10** (Structures

4

2) by using this method [9]. The Dewar benzene precursor was prepared by [2 + 2] cycloaddition of DMAD with a cyclobutadiene-AlCl$_3$ complex using the procedure developed by Hogeveen.

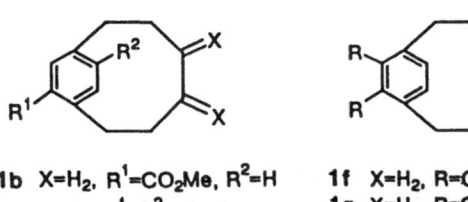

Scheme 3.

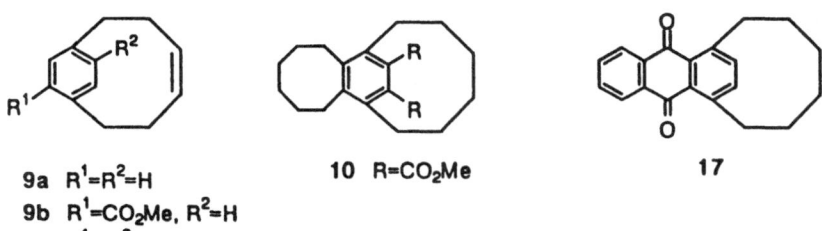

1b X=H$_2$, R^1=CO$_2$Me, R^2=H   **1f** X=H$_2$, R=CO$_2$Et
1c X=H$_2$, R^1=R^2=CO$_2$Me   **1g** X=H$_2$, R=CO$_2$Me
1d X=H$_2$, R^1=CO$_2$H, R^2=H   **1h** X=O, R=CO$_2$Me
1e X=O, R^1=CO$_2$Me, R^2=H   **1i** X=H$_2$, R=CH$_2$OH
   1j X=H$_2$, R=CH$_2$OAc
   1k X=H$_2$, R=CH$_2$OCONHC$_6$H$_{11}$

9a R^1=R^2=H   **10** R=CO$_2$Me   **17**
9b R^1=CO$_2$Me, R^2=H
9c R^1= R^2=CO$_2$Me

Structures 2

We also developed a method based on the cleavage of the central bond of the [6.2.2]propellane skeleton by way of carbocation rearrangement [5b, 10].

We had already found that oxidative decarboxylation of [6.2.2]propellane-carboxylic acid (**11**) with lead tetraacetate produced bridgehead alkene **12** with an acetoxyl group at the opposite bridgehead (Scheme 4) [11]. Subsequent elimination under thermal conditions gave dihydro[6]paracyclophane (**13**) [12]. We thought that a similar sequence of reactions with propellenecarboxylic acid (**14**) would give **1a**. Indeed, oxidative decarboxylation of **14** afforded **1a** directly as the major product along with the expected acetate **15**. The latter was cleanly converted to **1a** by treatment with *t*-BuOK. Following this procedure, a suffi-cient amount of the parent hydrocarbon **1a** became available for further study of its chemical reactivity.

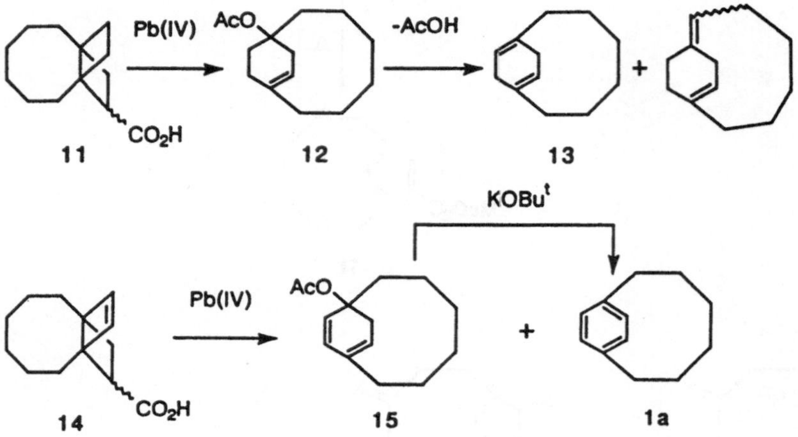

Scheme 4.

Another unique method for the synthesis of [6]paracyclophane derivatives was developed by Tochtermann, which was based on the deoxygenation of oxepinophane using the McMurry's reagent (Scheme 5) [13]. The oxepinophane

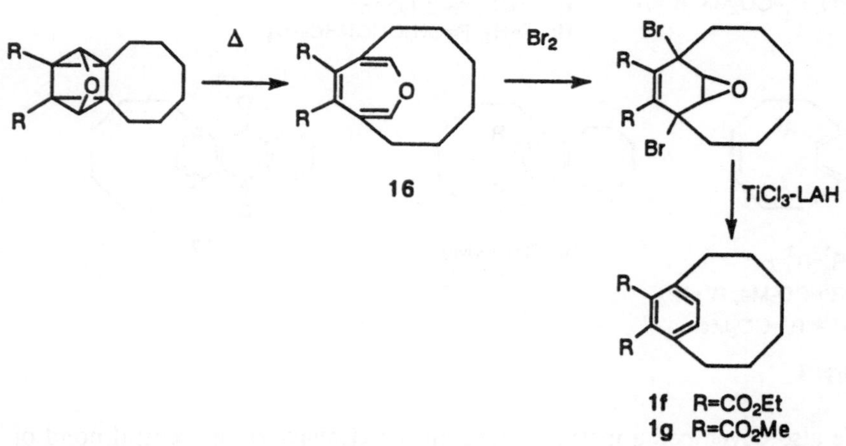

1f   R=CO₂Et
1g   R=CO₂Me

Scheme 5.

16 was prepared by (i) furan-acetylene [4 + 2] addition, (ii) intramolecular [2 + 2] addition, and (iii) thermal automerization of the oxaquadricyclane derivative. Bromination followed by reductive deoxygenation produced bis(ethoxycarbonyl)-substituted [6]paracyclophane (1f) in good yield. Methyl ester 1g, 3-keto derivative 1h [14], and optically active carboxylic acid 1d (Structures 2) [15] were also prepared by this procedure. Moreover, several hydroxymethyl based derivatives 1i–k [13b, 16] and anthraquinone derivative 17 [17] were prepared from 1g (Structures 2). Extensive studies on the structure and reactivity were undertaken for these compounds.

The Mo(CO)$_6$ mediated reaction of the bicyclic isoxazole 18 and DMAD afforded [6]pyridinophane (19) (Scheme 6) [18]. The reaction can be envisioned as a hybrid of the second (valence isomerization) and the fourth method (deoxygenation).

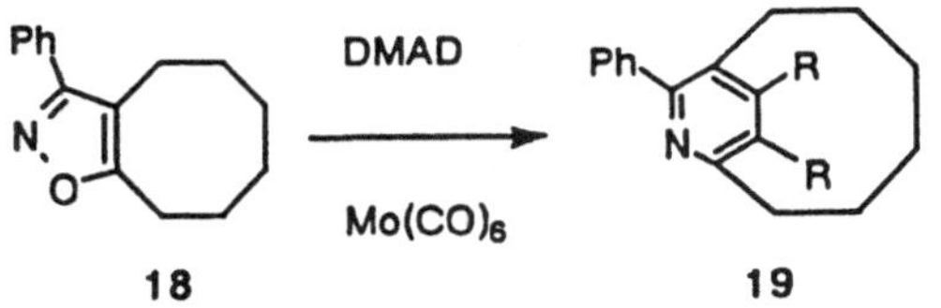

Scheme 6.

Only one of the methods used for the synthesis of [6]paracyclophanes was also successful in generation of the lower homologue, [5]paracyclophane (2a) (Structures 1), indicating clearly the limitation of these methods; thus FVP of lithium salt of tosylhydrazone 20 [3f], thermolysis of [5.2.2]propelladiene (21a) [19], oxidative decarboxylation of [5.2.2]propellenecarboxylic acid (22) [20], and reductive elimination of oxepinophane dibromide 23a and silyloxy-bridged derivative 23b (Structures 3) [21] all failed to give the [5]paracyclophanes. The only successful method proved to be the photochemical valence isomerization of Dewar benzene isomer 21a [22]. Thus irradiation (254 nm) of a solution of 21a in THF-d_8 at − 60 °C produced 2a in 7–8% yields (Scheme 7), which was unambiguously characterized by the ^{1}H NMR spectrum. The electronic spectrum (λ_{max} 280 and 330 nm) of 2a was in accordance with its structure. The low yield of 2a is perhaps due to the photochemical reversion to the starting material 21a. When the solution of 2a was warmed to room temperature, it decomposed immediately. Treatment of 2a at − 20 °C with TFA gave benzocycloheptene (24a) (Scheme 7), providing chemical evidence for the formation of 2a. Several mono-, di-, and tetrasubstituted derivatives like 21b–e were prepared either by propellanone ring contraction or the Hogeveen method and were irradiated with the hope that the steric and electronic effects would stabilize the corresponding [5]paracyclophanes 2b–e (Scheme 7) [23]. Although irradiation of 21b–e afforded 2b–e in 2–15% yields, attempts to isolate these compounds have so far been unsuccessful due to their lability.

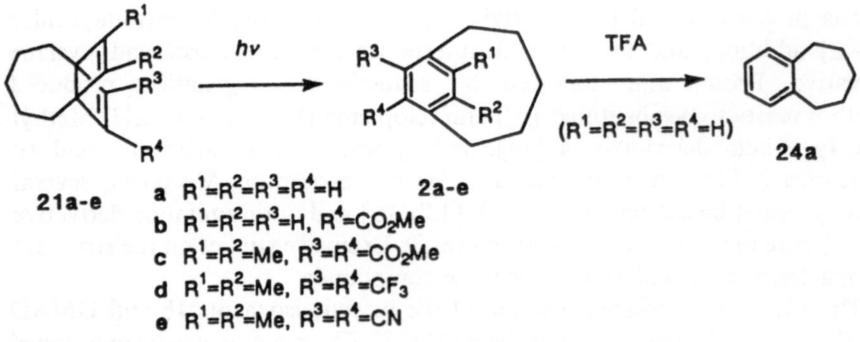
hν
TFA
$(R^1=R^2=R^3=R^4=H)$
21a-e
2a-e
24a
a $R^1=R^2=R^3=R^4=H$
b $R^1=R^2=R^3=H, R^4=CO_2Me$
c $R^1=R^2=Me, R^3=R^4=CO_2Me$
d $R^1=R^2=Me, R^3=R^4=CF_3$
e $R^1=R^2=Me, R^3=R^4=CN$

Scheme 7.

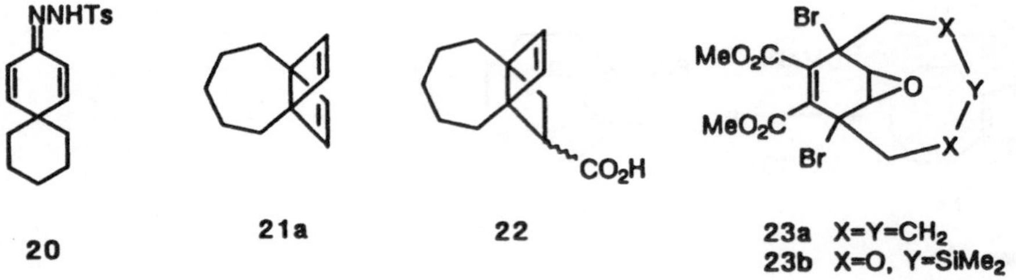
NNHTs
CO_2H
MeO_2C
MeO_2C
Br
Br
20
21a
22
23a X=Y=CH_2
23b X=O, Y=$SiMe_2$

Structures 3

Photochemical valence isomerization was also successfully applied to generation of [4]paracyclophane (**3a**). Tsuji and Nishida reported that irradiation of [4.2.2]propelladiene (**25a**) in ethanol at 77 K produced a reactive species possessing an electronic absorption at 340 nm which was assigned to **3a** (Scheme 8) [24]. The absorption disappeared when the rigid solution was warmed to − 100 °C indicating that **3a** decomposed at this temperature. Solution photolysis of **25a** in an alcohol gave ethers **26a–c** which were formed by nucleophilic 1,4-addition to **3a**. Similarly, irradiation of propelladienes **25b** and **25c** produced the cyclophane **3b** (λ_{max} 348, 425 nm) and **3c** (λ_{max} 340, 380 nm) which were intercepted by ethanol, giving the 1,4-adducts **27a–b** (Scheme 8). Bickelhaupt also reported independently that irradiation of **25a** in the presence of TFA or TFA and methanol gave the adducts **26a** and **26d** via acid actalyzed electrophilic addition to **3a**, indicating the intermediacy of **3a** [25]. Moreover, Tsuji and Nishida reported that photolysis of propellatetraenes **28a** at 77 K produced [4]paracyclopha-1,3-diene (**29a**; λ_{max} 347 nm) (Scheme 8) [26]. Irradiation of **28a** in the presence of cyclopentadiene gave the 2:1 adducts **30a** derived by two-fold Diels-Alder addition of cyclopentadiene to **29a**. Similarly, irradiation of the ester **28b** gave **30b** via **29b**. Since the butadienylene bridge of **29a** is shorter than the tetramethylene bridge of **3a**, cyclophane **29a** represents the most strained [*n*]paracyclophane known.

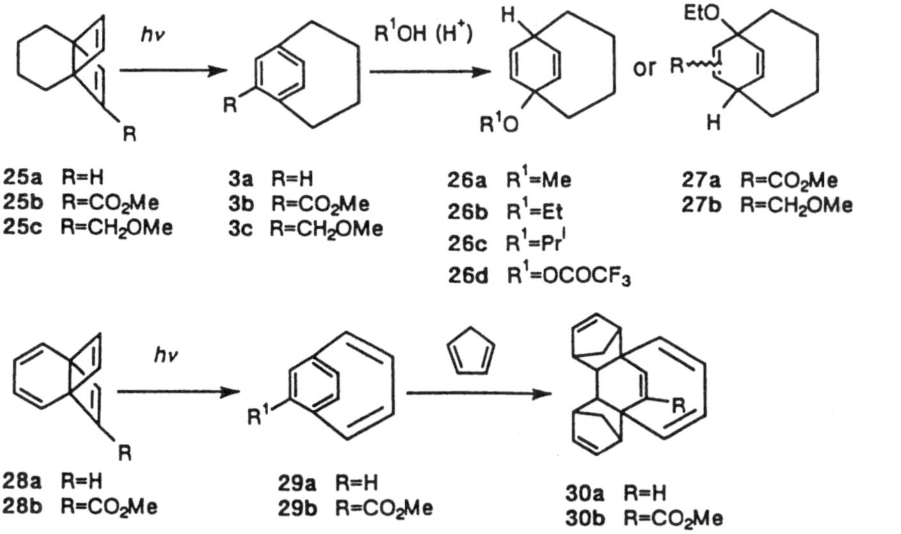

25a R=H
25b R=CO₂Me
25c R=CH₂OMe

3a R=H
3b R=CO₂Me
3c R=CH₂OMe

26a R¹=Me
26b R¹=Et
26c R¹=Prⁱ
26d R¹=OCOCF₃

27a R=CO₂Me
27b R=CH₂OMe

28a R=H
28b R=CO₂Me

29a R=H
29b R=CO₂Me

30a R=H
30b R=CO₂Me

Scheme 8.

2.2 Synthesis of Strained [n]Metacyclophanes

[n]Metacyclophanes of $n = 6-8$ were synthesized by several methods; these include thermal cleavage of the central bond of [n.3.1]propellanes [27], acid-catalyzed rearrangement of [n]paracyclophanes [5b, 10, 28], intramolecular diene-yne Diels-Alder reaction followed by dehydrogenation [29], and cyclo-addition of m-xylylene diradical with 1,3-dienes [30]. However, [5]meta-cyclophane (4a) and its derivatives were synthesized by Bickelhaupt only by the method based on the ring opening of [5.3.1]propellane skeleton under mild conditions (Scheme 9) [31]. Propellanes 32a–c were prepared by addition of dihalocarbene to the bicyclic olefin 31, which was obtained by addition of dichlorocarbene to 1,2-bis(methylene)cycloheptane followed by thermal vinyl-cyclopropane rearrangement. Treatment of dichloropropellane 32a with t-BuOK gave the parent hydrocarbon 4a. Silver-(I)-induced elimination/ring opening of tetrahalopropellanes 32b and 32c afforded metacyclophanes 4b and 4c, respectively (Scheme 9). This procedure opened an efficient route to [5]meta-cyclophanes and extensive studies on the structure and reactivity of this system were undertaken by Bickelhaupt's group.

In contrast to the [5.3.1]propellanes, reaction of chloro[4.3.1]propellene 33 or dihalo[4.3.1]propellane 34 with t-BuOK yielded Dewar benzene isomer 35 of [4]metacyclophane (5) instead of 5 itself (Scheme 10) [32]. Attempts to synthe-size and characterize 5 by either thermal or photochemical valence isomeriz-ation of 35 did not meet with success, although its intermediacy was indicated [33]. Photolysis of 35 at − 50 °C did not give 5 but led to the formation of prismane isomer 36 (Scheme 10). In contrast, thermolysis of 35 in a sealed

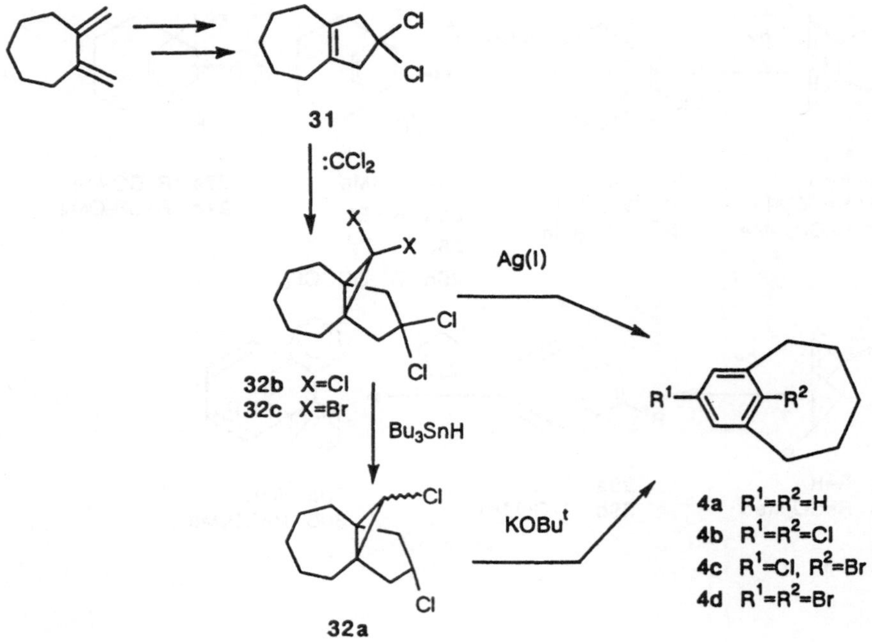

Scheme 9.

ampoule at 150–175 °C gave the cage dimers **37a** and **37b** as the major products, which were derived from initially formed cyclophane **5** by [4 + 2] dimerization and the subsequent intramolecular [4 + 2] cyclization (Scheme 11). Thermolysis of **35** at higher temperatures (200 °C) gave products containing [4.4]metacyclophane **38a** and [4.4]paracyclophane **38b** as the major components which were formed by two-fold retro [4 + 2] cleavage of **37a** and **37b**, respectively. FVP of **35** at 400–500 °C, however, afforded tetralin (**39a**) and methylindanes **40a–b** (Scheme 11). Formation of the products was proved by thermal isomerization of **5** using labeling experiments.

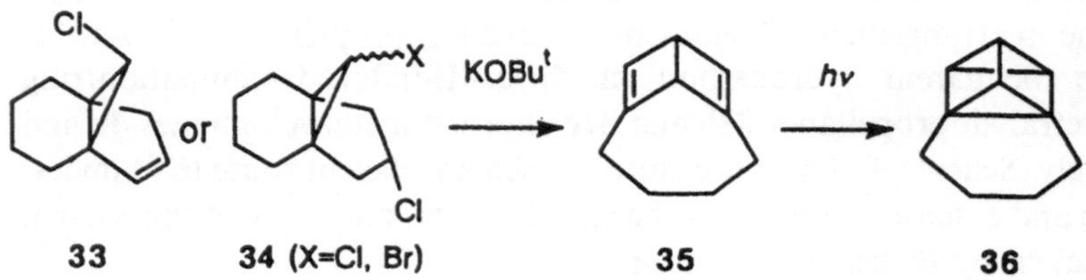

Scheme 10.

2.3 Synthesis of Strained [n]Cyclophanes of Condensed Benzenoid Aromatics

1,4-Bridged [n]cyclophanes of condensed benzenoid aromatics with $n \geq 8$ were prepared by the ordinary methods of cyclophane synthesis. These include [8]

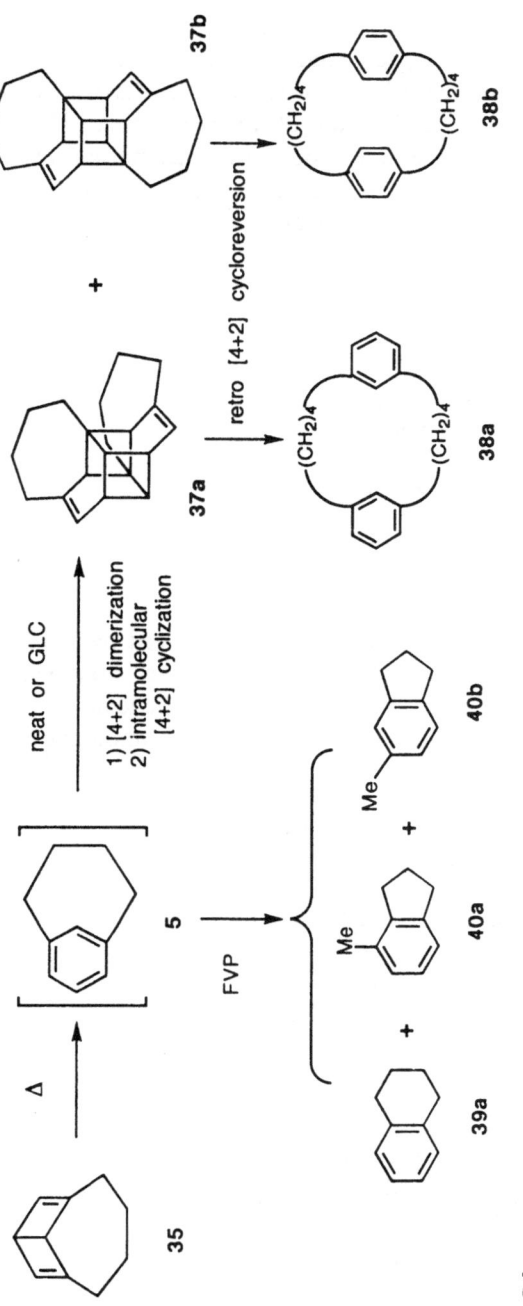

Scheme 11.

(1,4)naphthalenophane (**41a**) [34] and its diketo derivative **41b** [35], [8]
(9, 10)anthracenophane-3,6-dione (**42a**) [36], and 2,7-dithia[8](9,10)-anthra-
cenophane (**42b**) [37] (Structures 4). Recently, a new route to 2,7-diketo derivative **42c** (Structures 4) was developed by Fukazawa [38]. Some of these [8]cyclo-
phanes are reported to be fairly unstable even though their aromatic rings are
only modestly deformed from planarity. Consequently, it may well be anticipated that most of the common methods will not be applicable to the synthesis of
the lower homologues with $n \leq 6$.

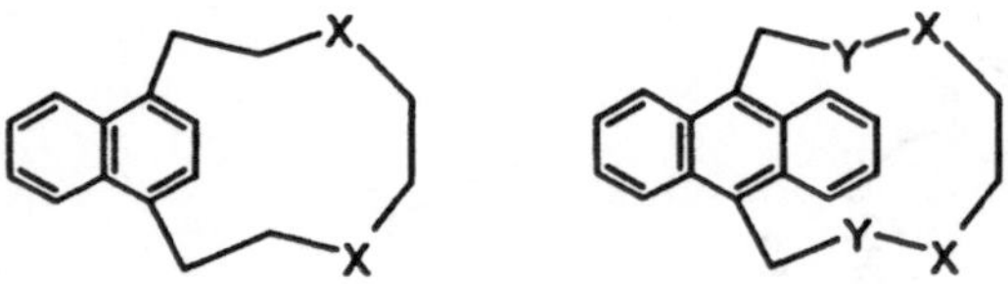

41a X=CH$_2$
41b X=CO

42a X=CO, Y=CH$_2$
42b X=CH$_2$, Y=S
42c X=CH$_2$, Y=CO

Structures 4

Thus [7](1,4)naphthalenophane derivative **45** was synthesized by Tochter-
mann by deoxygenation of oxepinophane **44** which was in turn obtained by
Diels-Alder reaction of the furooxepinophane **43** with DMAD (Scheme 12) [39].
This method has not been tested for the synthesis of a lower homologue than
$n = 6$.

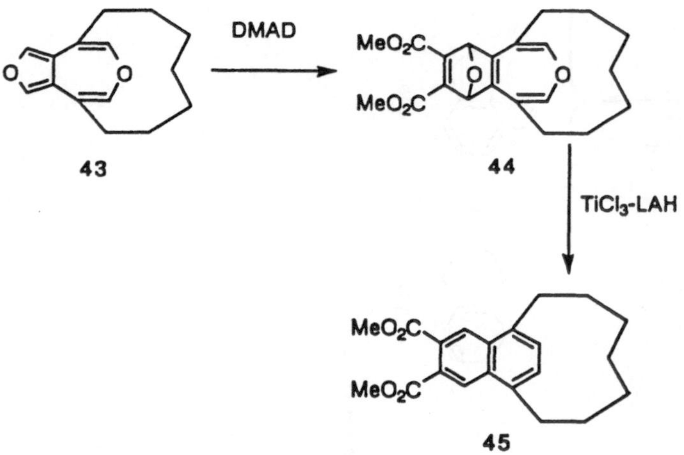

43   **44**

45

Scheme 12.

We devised routes to [6](1,4)naphthalenophanes **55a–b** and [6](1,4)an-
thracenophanes **56a–b** based on the same method developed for the synthesis of
[6]paracyclophanes; the cleavage of the central bond of the propellane skeleton

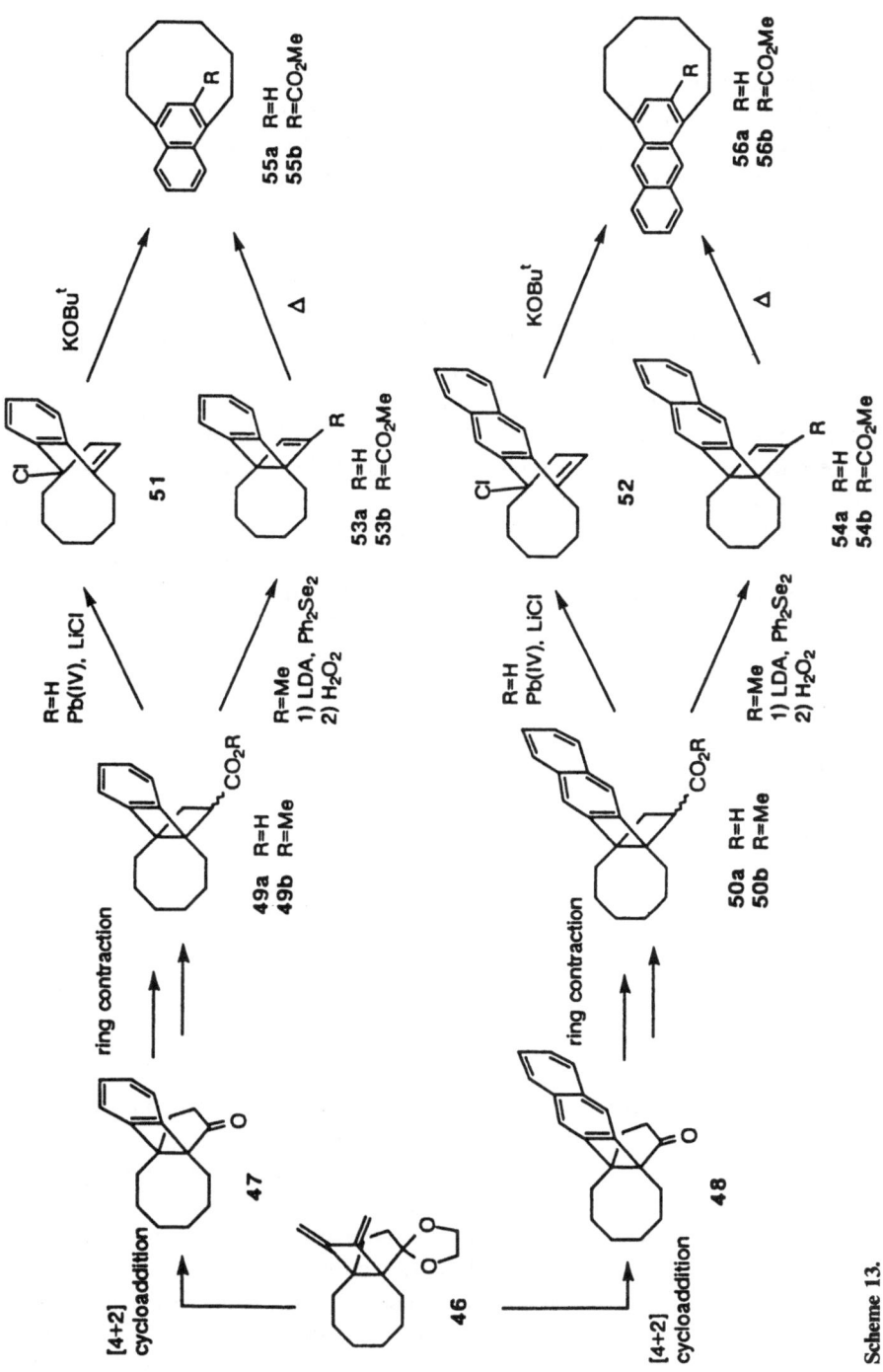

Scheme 13.

Yoshito Tobe

by either carbocation rearrangement or valence isomerization of Dewar type isomers, respectively (Scheme 13) [40]. To this end, the benzo- and naphtho-fused propellanones 47 and 48 were prepared from the propelladiene 46 using Diels-Alder reactions. Ring contraction of 47 and 48 gave the acids 49a and 50a or the esters 49b and 50b, respectively. Oxidative decarboxylation of 49a and 50a under the conditions used for the synthesis of the benzologue 1a did not give the desired products probably because of the lability of the products under the reaction conditions. However, chlorinative decarboxylation of 49a and 50a in the presence of LiCl in DMSO at room temperature furnished the bridgehead chlorides 51 and 52, respectively. Treatment of the chlorides with t-BuOK afforded the unstable cyclophanes 55a and 56a, respectively. Moreover, the esters 49b and 50b were converted to the Dewar type isomers 53b and 54b. Thermal valence isomerization of the Dewar isomers proceeded smoothly to afford the mono substituted derivatives 55b and 56b (Scheme 13).

Recently, we developed a second route to 55a and 56a (Scheme 14) [41]. Bromocyclophane 1l was treated with a strong base to generate a benzyne intermediate, which was trapped with furan or isobenzofuran to afford [4 + 2] adducts 57 (two isomers; 6:1 ratio) and 58 (two isomers; 2:1 ratio), respectively. Deoxygenation of 57 and 58 with a low valent titanium reagent prepared from $TiCl_4$, $LiAlH_4$, and triethylamine gave 55a and 56a, respectively.

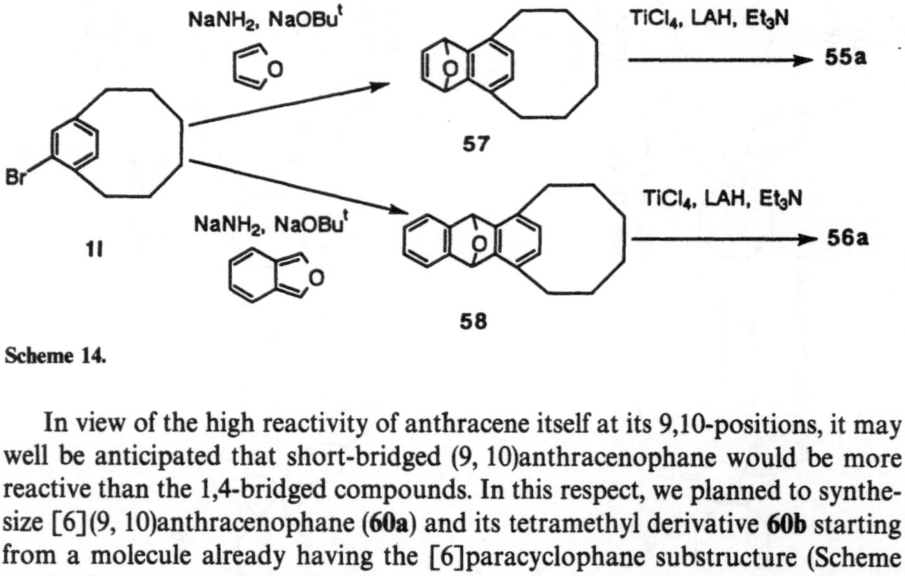

Scheme 14.

In view of the high reactivity of anthracene itself at its 9,10-positions, it may well be anticipated that short-bridged (9, 10)anthracenophane would be more reactive than the 1,4-bridged compounds. In this respect, we planned to synthe-size [6](9, 10)anthracenophane (60a) and its tetramethyl derivative 60b starting from a molecule already having the [6]paracyclophane substructure (Scheme 15) [42]. Thus treatment of dibromocyclophane 1m with a strong base in the presence of excess furan or 2,5-dimethylfuran gave the bis[4 + 2] adducts 59a (three isomers; 3:2:1) and 59b (three isomers; 1:2:4), respectively. Reductive deoxygenation of 59a yielded dihydroanthracenophane 61a and the methyl-enedihydroanthracene 62a in a ratio of 2:1. The desired cyclophane 60a, however, was not detected. On the other hand, similar reaction of 59b gave anthracenophane 60b as the major product together with a small amount of

14

its isomer **62b** and the dihydro derivative **61b** (11:1:2). A pure sample of **60b** (mp 161–162 °C) obtained by recrystallization was fairly stable to air but was very sensitive to acid.

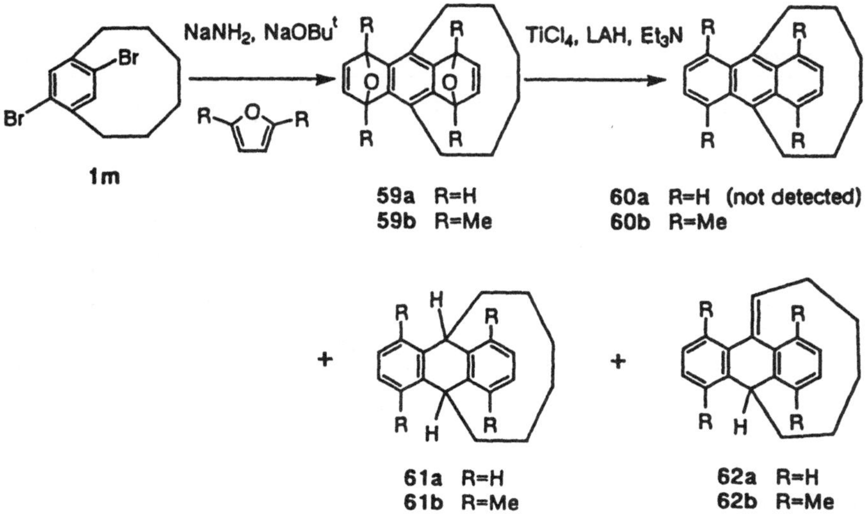
NaNH2, NaOBu^t
TiCl4, LAH, Et3N
1m
59a R=H
59b R=Me
60a R=H (not detected)
60b R=Me
61a R=H
61b R=Me
62a R=H
62b R=Me

Scheme 15.

As to the 1,3-bridged [n]cyclophane of condensed benzenoid aromatics, 2-chloro[5](1, 3)naphthalenophane (**64**) is the smallest representative. This compound was synthesized by Reese by silver-(I)-induced ring opening of benzo[5.3.1]propellene **63** (Scheme 16) [43].

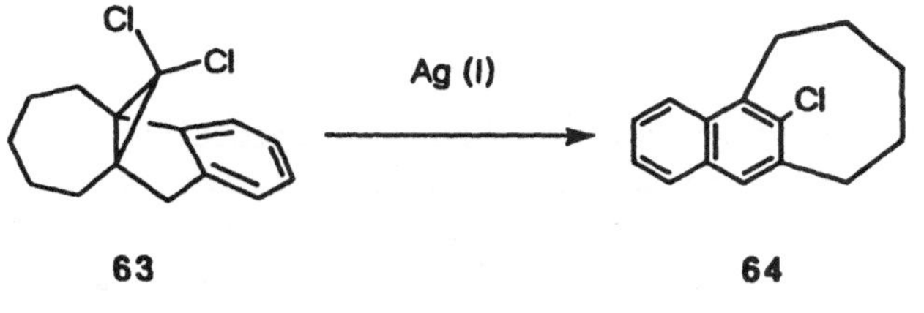
Ag (I)
63
64

Scheme 16.

3 Geometry and Aromaticity of Strained [n]Cyclophanes

3.1 Theoretical Studies on Geometry and Aromaticity of Strained [n]Cyclophanes

Several theoretical calculations have already been done for strained [n]paracyclophanes by means of semi-empirical molecular orbital calculations and molecular mechanics calculations. However, recent advance in the synthetic studies of these molecules led to reinvestigation of the theoretical approach based on high level of ab initio calculations [44–47]. The calculated geometries of paracyclophanes **1a**, **2a** and **3a** (Structures 1) are summarized in Table 1.

Table 1. Ab initio calculated geometries of strained [n]paracyclophanes **1a**, **2a**, and **3a** (Structures 1)

| Compd | Symmetry | Basis Set | Deformation Angle[a] | | Aromatic Bond Length (Å) | Ref. |
			α (deg)	β (deg)		
1a	C_2	STO-3G	17.2	21.7	1.382–1.394	44c
		6-31G	18.8	–	1.39	47
		Double ζ	18.6	20.7	1.392–1.399	44c
2a	C_s	STO-3G	23.4	29.8	1.365–1.412	44a
		6-31G	23.5	–	1.38–1.40	47
		Double ζ	23.7	28.6	1.384–1.409	44b
3a	C_2	STO-3G	28.3	39.5		45
		Double ζ	29.7	38.2	1.382–1.396	46

[a] See Fig. 1.

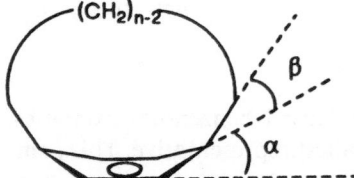

Fig. 1. Schematic representation of the deformation angles of [n]paracyclophanes

The calculated geometries for [6]paracyclophane (**1a**), particularly those derived by using the double ζ basis set, are in much better agreement with the experimental values than those obtained by previous works. The bond length alternation of the cyclophanes is small; within 0.007 Å for **1a**, 0.025 Å for **2a**, and 0.014 Å for **3a**. The largest deviation in **2a** is probably due to the C_S symmetry of the molecule. These indicate that, despite the large deformation of the aromatic ring from planarity, **1a**–**3a** should be regarded as aromatic. On the other hand, the resonance energies of **1a** and **2a**, estimated by Schaeffer from a homodesmetic reaction, are approximately 54 and 78 kcal/mol less than that of benzene. This would lead to the conclusion that **1a** and **2a** are not aromatic in the thermodynamic sense, if one does not take the geometrical strain into account [44b]. From the fact that the SCF optimization was not suitable for the calculation of the energy difference between **3a** and its Dewar benzene isomer **25a** (s. Scheme 8), Grimme suggested that **3a** could be classified as the borderline case between a closed shell molecule and a diradical like species [46].

For several [6]paracyclophanes with unsaturated bridges like **9a** (Structures 2) and its benzologue **65a** (Structure 5), MNDO calculations were undertaken and the energies were compared with those of the corresponding Dewar type valence isomers [48].

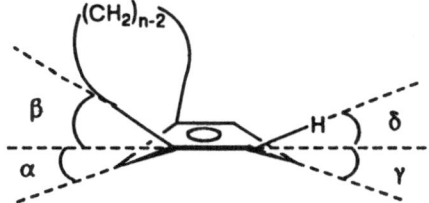

65a R=H
65b R=CO$_2$Me

Structure 5

MNDO calculations for [n]metacyclophanes **4a** and **5** (Structures 1) and STO-3G ab initio calculations for **4a** were done by Jenneskens [49]. The structural parameters are listed in Table 2. The STO-3G geometries of **4a** agree reasonably with those experimentally determined for the 8,11-dichloro derivative **4b** (s. Scheme 9). The small deviation in the calculated bond lengths of **4a** and **5** indicates that these should be regarded as aromatic. Aromatic delocalization energies of **4a** and **5** were estimated by comparing of the energies of delocalized benzene and localized 1,3,5-cyclohexatriene whose bent angles were the same as those of **4a** and **5**, respectively. The fact that the delocalization energies of bent benzenes are similar to that of planar benzene leads to the conclusion that bond localization is unfavorable even in the case of highly bent **5**.

Fig. 2. Schematic representation of the deformation angles of [n]metacyclophanes

Table 2. MNDO and ab initio calculated geometries of strained [n]metacyclophanes **4a** and **5** (Structures 1)

| Compd | Method | Deformation Angle[a] | | | | Aromatic Bond | |
		α (deg)	β (deg)	γ (deg)	δ (deg)	Length (Å)	Ref.
4a	MNDO	32.5	43.1	9.2	7.7	1.409–1.416	49a
	STO-3G	22.2	40.9	11.1	17.9	1.381–1.392	49b
5	MNDO	40.6	46.1	12.0	12.0	1.407–1.421	49a

[a] See Fig. 2.

Yoshito Tobe

In order to obtain insight into the reactivity of [n]cyclophanes of condensed benzenoid aromatics, we undertook semi-empirical molecular orbital calculations for 1,4- and 1,3-bridged [6]naphthalenophanes **55a** and **67**, [6]anthracenophanes **56a** and **68**, and 9,10-anthracenophanes **60a** and **60b** (s. Schemes 13 and 15, Structures 6) [42, 50]. The calculated deformation angles are listed in Table 3. Comparison of these geometries with those of the corresponding benzenophanes **1a** and **66** (Structures 1 and 6) indicates clearly that the bending angles, especially the deformation angle α, increase as the size of the aromatic ring increases due to the greater flexibility of the larger aromatic rings. Particularly noteworthy is that the π bond orders of the aromatic bond *a* of

Table 3. AM1 and PM3 calculated geometries of strained [6]cyclophanes **1a** and **66** (Structures 1 and 6) and their condensed benzenoid analogues **55a, 56a, 60a, 60b, 67** and **68** (Schemes 13 and 15, Structures 6)

Compd	Hamiltonian	SE[b](kcal/mol)	α(deg)	β(deg)	γ(deg)	Ref.
1a	PM3	36.2	23.2	16.8	–	50
55a	PM3	37.5	25.2	15.8	–	50
56a	PM3	41.2	25.7	15.5	–	50
	AM1	39.4	25.4	16.0	–	42
60a	AM1	41.2	25.6	16.1	–	42
60b	AM1	–	28.8	14.8	–	42
66	PM3	23.8	19.5	20.5	7.6	50
67	PM3	25.0	20.3	18.6, 23.0	8.8	50
68	PM3	29.4	20.8	19.2, 22.8	9.6	50

Deformation Angle[a]

[a] See Figs. 1 and 2
[b] Strain energies were estimated using Benson's standard group increments.

66   67   68

bond a R   bond a R   Me Me Me

bond b   bond b

R   R   Me Me Me

69a R=Me   70a R=Me   71
69b R=Et   70b R=Et

Structures 6

18

(1,4)naphthaleno- and (1,4)anthracenophanes **55a** and **56a** are larger than those of the reference compounds **69a** and **70a**, despite the fact that an overlap of π orbitals is less favorable. As a result, the unbridged aromatic rings of **55a** and **56a** have more 6π and 10π like character, respectively, than those of the corresponding references. These observations are in accordance with their extraordinary reactivities.

3.2 Experimental Studies on Geometry and Aromaticity of Strained [n]Cyclophanes

3.2.1 Spectral Properties

The longest-wavelength absorption bands in the electronic spectra of strained [n]para- and [n]metacyclophanes and their condensed benzenoid analogues are summarized in Table 4 together with their reference compounds. It can be seen that, as the length of the bridge n decreases, the [n]cyclophanes exhibit remarkable bathochromic shift. This is mainly due to decrease of the HOMO/LUMO gap as a result of destabilization of the former and stabilization of the latter.

Table 4. Longest wavelength absorption band of strained [n]paracyclophanes **1a**, **9a**, **2a**, **3a**, and **29a**, [5]metacyclophane **4a**, [6]naphthalenophane **55a**, [6]anthacenophanes **56a** and **60b** (Structures 1 and 2); (Schemes 8, 13 and 15) and their reference compounds

Compd	Solvent	λ_{max}(nm)(log ε)	Ref.
1a	ethanol	296(2.8)	1
9a	hexane	310(2.5)	8 a, c
2a	hexane	330(–)	22
3a	ethanol	340(–)	24
29a	EPA[a]	347(–)	26
p-diethylbenzene	hexane	266(2.5)	–
4a	cyclohexane	307(–)	31d
m-xylene	cyclohexane	265(2.5)	–
55a	cyclohexane	314(3.5)	40b
1,4-diethylnaphthalene (**69b**)	cyclohexane	289(3.8)	40b
56a	cyclohexane	385(3.6)	40b
1,4-diethylanthracene (**70b**)	cyclohexane	365(3.7)	40b
60b	cyclohexane	455(3.6)	42
1,4,5,8,9,10-hexamethylanthracene (**71**)	heptane	425(3.7)	51

[a] ether: pentane: ethanol = 5:5:2.

Thorough assignment of the NMR signals of [6]paracyclophane derivative **1g** (Structures 2) was undertaken by Tochtermann and Günther by means of the 2D NMR techniques [52]. The ^{1}H NMR spectra of [6]paracyclophanes exhibit dynamic behavior due to slow flipping of the bridge on the NMR time scale (Scheme 17). The barrier for the bridge rotation of **1g** was determined by the line-shape analysis to be 14.0 kcal/mol (298 K). The barriers for the parent 1a as well as other [6]paracyclophane derivatives, estimated from the coalescence temperature and the frequency difference between the two sites, are in the range of 13–14 kcal/mol, indicating little effect of the substituents [5b, 6, 7, 14, 16, 53]. However, we found that the barrier for the benzyne-furan adducts 59a and 59b (s. Scheme 15) vary from 12 to 15 kcal/mol depending on the orientation of the oxygen relative to the methylene bridge [41]. The barrier for flipping the bridge of cyclophene 9a (Structures 2), having a *cis* double bond in the bridge, is much higher (23 kcal/mol); it was determined from the *syn/anti* isomerization rates of the mono-ester derivative 9b (Structures 2) [8b, c]. Moreover, benzannelation enhanced the barrier substantially (> 26 kcal/mol) since no *syn/anti* isomerization was observed for the [2.2]orthoparacyclophane 65b (Structure 5) before it decomposed [54]. Similar effect of benzannelation was observed for [2.2]orthometacyclophane (72) (Structures 7) whose conformational barrier ($20 < \varDelta G^{\ddagger} < 24$ kcal/mol) [55] was considerably larger than that of [6]metacycloph-3-ene (73) (Structures 7) ($\varDelta G^{\ddagger}(343$ K$) = 16.5$ kcal/mol) [30]. As to [5]paracyclophane (2a) (Structures 1), the barrier for the flipping of the bridge was determined to be 14.3 kcal/mol (224 K) [22].

1 a

Scheme 17.

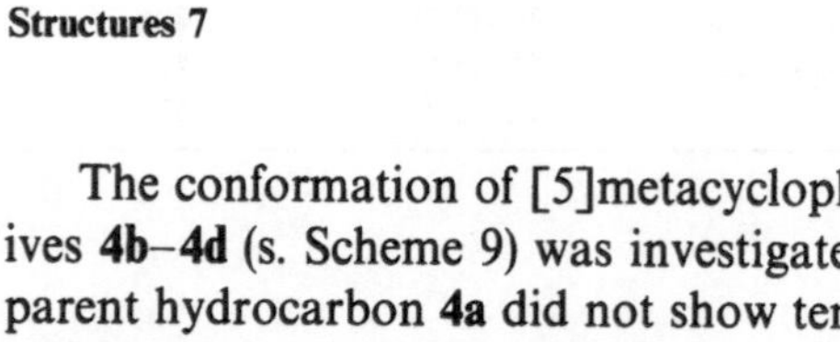

72   73

Structures 7

The conformation of [5]metacyclophane (4a) (Structures 1) and its derivatives 4b–4d (s. Scheme 9) was investigated by Bickelhaupt [31d, 56]. While the parent hydrocarbon 4a did not show temperature dependence in the ^{1}H NMR, 4b–4d exhibited dynamic behavior due to the conformational change between the two conformers A and B (Scheme 18). The conformer A predominates over B by 6–8 times; the ratio is dependent on the substituent. The barrier for the flipping was estimated to be 13.2–13.4 kcal/mol (296 K) from the line-shape analysis and the spin saturation transfer experiments.

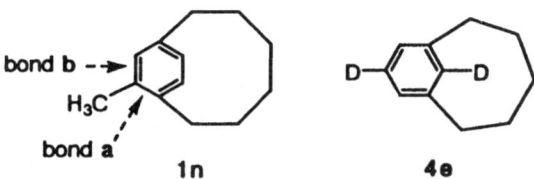

conformer A   conformer B

Scheme 18.

The conformational barrier for the 1,4-naphthalenophane **55a** (13.6 kcal/mol (273 K)) and anthracenophane **56a** (s. Scheme 13) (13.4 kcal/mol (278 K)) are almost similar to, but slightly smaller than, that of **1a** (Scheme 17) (13.9 kcal/mol (278 K)) [40b]. On the other hand, the barrier for the tetramethyl (9, 10)anthracenophane **60b** (s. Scheme 15) is substantially smaller (9.5 kcal/mol) [42]. This is probably due to destabilization of the ground state conformation because of steric repulsion between the methyl groups and the benzyl methylenes.

The ^4J (H–H) coupling constant between the methyl and aromatic protons of 8-methyl[6]paracyclophane (**1n**) (Structures 8) was determined (− 0.61 Hz). It was slightly smaller than the less strained paracyclophanes (− 0.66 Hz) [57]. Similarly, the ^3J (H–H) coupling constant of the aromatic protons of **1n** (7.27 Hz) is smaller than the others (7.62–7.85 Hz), though it is a less reliable measure of the π bond order. These indicate that the π bond order of bond *b* of **1n** is smaller than normal. STO-3G calculations for the bent benzene models, however, predict that the π bond order of the bond *b* does not change even when the bent angle increases up to 30°. The ^1J (C–H) coupling constants of **1a** (Structures 1) (165.3 and 159.8 Hz at − 40 °C) are larger than that of *p*-diethylbenzene (156.4 Hz), indicating that the *s* character of the aromatic C–H bond of **1a** is larger than that of the reference compound [58]. On the other hand, the vibrational frequency analysis of the ab initio calculations predicts that the *s* character of the aromatic C–H bond of **1a** is smaller than that of *p*-dideuterobenzene [44c]. These contradictions remain to be clarified.

bond b →

H₃C

bond a

1 n

D—⬡—D **4 e**

Structures 8

The magnetic susceptibility of dideutero[5]metacyclophane (**4e**) (Structures 8) was determined by means of quadrupolar deuterium coupling at very high field [59]. The susceptibility of **4e** is similar (or even lower than) to that of appropriate reference compounds, indicating that the [5]paracyclophane system is fully aromatic in spite of its highly deformed geometry of the benzene ring.

3.2.2 X-ray Crystallographic Analyses

So far, X-ray crystallographic structure analyses have been done for four [6]paracyclophane derivatives **1c** [6], **1d** [5], **1f** (Structures 2) [13b], and **74** (Structures 9) [50] by Tochtermann and us. As to [n]paracyclophanes of condensed benzenoid aromatics, the structures of [7](1,4)naphthalenophane derivative **45** (s. Scheme 12) [39], [6](1,4)anthracenophane (**56a**) (s. Scheme 13) [40b], and [6](9,10)anthracenophane derivative **60b** (s. Scheme 15) [42] have been determined. ORTEP drawings of **1c**, **56a**, and **60b** are shown in Fig. 3. As can be seen from the top view (Fig. 3a), the bridge methylene is of C_2 symmetry. The side views (Fig. 3b and 3c) illustrate remarkable deformation of the aromatic ring into a boat shape. The selected structural parameters of these compounds are listed in Table 5. It can been seen that the bending angle (α) of the *para* carbon of [6]paracyclophane system is approximately 19.5° and that of the benzyl carbon (β) is 20°. The recent molecular orbital calculations reproduce these features well (Table 1). The angle α of the bridged cyclophane **74** (Structures 9) is somewhat smaller while its β is larger than the others probably because of the steric effect of the ethano bridge. The observed bond distances vary much more than the calculated mainly because of the electronic effect of the substituents. The bending angle of the [6]paracycloph-3-ene derivative **9c** (Structures 2) is the largest of the known [n]paracyclophanes.

The deformation angle α of 1,4-bridged anthracene **56a** (s. Scheme 13) is slightly larger than that of [6]paracyclophanes while the angle β is almost the same. Furthermore, the deformation angle α of the 9,10-bridged anthracene **60b** (s. Scheme 15) is substantially larger than that of **56a**, although the angle β is slightly smaller. This may be partly due to the greater flexibility of the central anthracene ring than that of the side rings, but mainly due to the steric repulsion between the methyl groups and the benzyl methylenes, since the molecular orbital calculations predict similar degree of deformation in **56a** and **60a** (Table 3). The central ring of **60b** represents the most highly deformed boat shaped aromatic ring known.

Only a few structural studies have been done for strained [n]metacyclophanes. Selected parameters for the [5]metacyclophane **4b** (s. Scheme 9) [60], oxa[6]metacyclophane **75** [29], and (1,3)naphthalenophane derivative **76** (Structures 9) [50] are summarized in Table 6. The deformation in **4b** is remarkably larger than those in **75** and **76**. The total bent angle ($\alpha + \gamma$) of the aromatic ring of **4b** (38.8°) is almost the same as that of the [6]paracyclophanes system. Moreover, the large deformation angle β (48°) is characteristic in the metacyclophane. The bending angle (α) of the naphthalenophane **76** is substantially larger than that of the benzenophane **75** of the same bridge carbon number, probably due to the greater flexibility of the naphthalene ring compared to benzene.

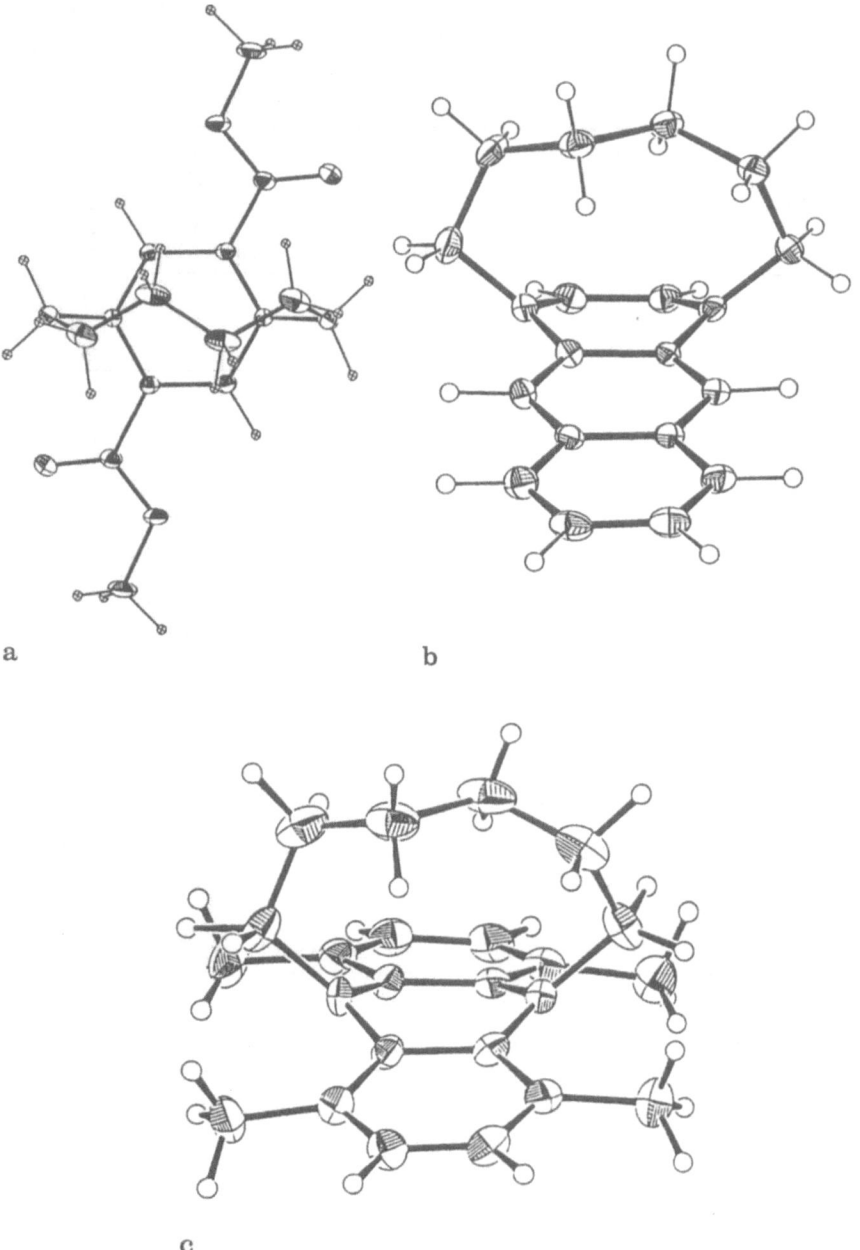

a   b

                          c

Fig. 3. ORTEP drawings of molecular structures of **1c** (a: top view), **56a** (b: side view), and **60b** (c: side view) (Structures 1, Schemes 13 and 15)

Table 5. Crystal structure geometries of [6]paracyclophanes **1c, 1d, 1g, 74,** and **9c** (Structures 2 and 9), and benzenoid aromatic analogues **45, 56a,** and **60b** (Schemes 12, 13 and 15)

| Compd | Deformation Angle[a] | | Aromatic Bond Length | | Ref. |
	α (deg)	β (deg)	Bond a (Å)	Bond b (Å)	
1c	19.4	20.2	1.395	1.388–1.414	6
1d	20.1, 20.9	18.1, 19.2	1.376–1.399	1.385–1.393	5
1g	19.4, 19.5	18.6, 21.2	1.370–1.395	1.385–1.393	13b
74	17.6, 18.5	21.0, 23.1	1.37–1.399	1.37–1.39	50
9c	20.5	24.1	1.404	1.388–1.413	8c
45	13.6, 15.9	13.0, 13.8	–	–	39
56a	20.5, 21.5	19.4, 19.6	–	–	40b
60b	23.8, 25.5	18.4, 18.7	–	–	42

[a] See Fig. 1.

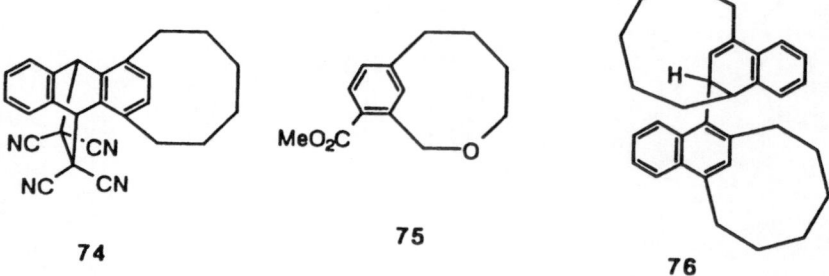

74

75

76

Structures 9

Table 6. Crystal structure geometries of [5]metacyclophane **4b**, oxa[6]metacyclophane **75**, and [6](1, 3)naphthalenophane **76** (Scheme 9, Structures 9)

| Compd | Deformation Angle[a] | | | Aromatic Bond Length (Å) | Ref. |
	α (deg)	β (deg)	γ (deg)		
4b	26.8	48.0	12.0	1.389–1.400	60
75	17.0	22.9	6.4	–	29
76	25.3	13.2, 18.8	4.6	–	50

[a] See Fig. 2.

4 Reactivity of Strained [n]Cyclophanes

Strained [n]cyclophanes exhibit enhanced reactivity compared to unstrained aromatics. Frequently they show a variety of extraordinary reactivities which are not encountered in the chemistry of aromatic compounds. In general, these involve isomerizations induced thermally, photochemically, or by acid, electrophilic addition reaction, and nucleophilic substitution and addition reactions.

4.1 Thermal and Photochemical Reactions

Vapor phase thermolysis of [6]paracyclophane (**1a**) (Structures 1) yielded spiro triene **77** (Structures 10) as the major product via homolytic cleavage of one of the benzylic bonds [5b, 10]. FVP of [5.2.2]propelladiene (**21a**) (Structures 3) afforded spiro compound **78** (Structures 10) by similar homolysis of intermediate [5]paracyclophane (**2a**) (Structures 1) [61]. In contrast, the reverse reaction, FVP of spiro trienes, was successfully used for the preparation of [n]paracyclophanes with n = 7 and 8 [62]. Pyrolysis of spiro tetraene **79**, however, did not give [7]paracycloph-3-ene (**80**) (Structures 10) [63]. Evidently, [7]paracyclophane is the borderline case with regard to the thermodynamic stabilities between the bridged aromatic compound and the spiro triene isomer.

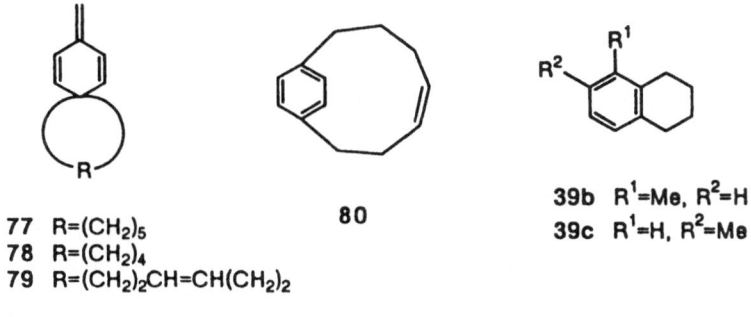

77 R=(CH$_2$)$_5$
78 R=(CH$_2$)$_4$
79 R=(CH$_2$)$_2$CH=CH(CH$_2$)$_2$

80

39b R^1=Me, R^2=H
39c R^1=H, R^2=Me

Structures 10

FVP of [5]metacyclophane (**4a**) (Structures 1) at 600 °C yielded methyltetralins **39b** and **39c** (Structures 10) as the major products via homolysis of the benzylic bond [33c]. On the other hand, FVP of the Dewar isomer **35** produced retralin **39a** and methylindans **40a–b** via [4]metacyclophane (**5**) (Scheme 11). Pyrolysis of neat **35** gave dimers **37a–b** derived by [4 + 2]dimerization of intermediate **5** (Scheme 11) [33a].

Photolysis of **1a** produced the Dewar benzene isomer **8a** (s. Scheme 2) [4]. It should be emphasized that similar photoisomerization of the higher homologues (n = 7, 8) did not take place but gave merely polymers on irradiation [4, 28b]. Since photochemical cycloreversion of **8a** to **1a** took place, further irradiation led to a photostationary state composed of **1a** and **8a**. Similar photoisomerization of the derivatives **1b**, **1f**, **1i**, **1j**, **1k**, and **10** (Structures 2) was also reported [5, 9, 13b, 16, 64]. Prolonged irradiation of **1b**, **1f**, **1k**, and **10** produced the corresponding prismane type isomers **81a–d** (Structures 11). Thermal cycloreversion of the Dewar type isomers **8a–d** and those of **82a–c** (Structures 11) took place at 40–50 °C, giving cleanly the corresponding aromatic compounds. The activation energies of the isomerization vary from 22 to 29 kcal/mol, depending on the substituents. Photocycloreversion of [5]paracyclophane (**1b**) and its derivatives to the corresponding Dewar isomers **21a–e** was also assumed to occur, since irradiation of the Dewar isomers led to photostationary states [22, 23].

8a $R^1=R^2=R^3=H$

8b $R^1=CO_2Me$, $R^2=R^3=H$

8c $R^1=R^3=CO_2Me$, $R^2=H$

8d $R^1=H$, $R^2=R^3=$
    $CH_2OCONHC_6H_{11}$

81a $R^1=R^2=R^3=H$, $R^4=CO_2Me$

81b $R^1=R^2=H$, $R^3=R^4=CO_2Et$

81c $R^1=R^2=H$, $R^3=R^4=$
     $CH_2OCONHC_6H_{11}$

81d $R^1=R^2=(CH_2)_6$, $R^3=R^4=CO_2Me$

82a $R^1=R^2=H$

82b $R^1=H$, $R^2=CO_2Me$

82c $R^1=R^2=CO_2Me$

Structures 11

On the other hand, Miki reported that photolysis of the anthraquinone **17** (Structures 2) did not give its Dewar isomer but the methylenedihydro isomer **83** (Structures 12) which was formed by intramolecular hydrogen abstraction from the benzyl methylene by the triplet-excited carbonyl [65].

In contrast to the paracyclophanes, photolysis of [5]metacyclophane (**4a**) (Structures 1) gave benzocycloheptene **24a** (Structures 12) through isomerization to a benzvalene type isomer which isomerized thermally to the final product [66]. The mechanism of the isomerization was confirmed by the ^{13}C labeling experiment. Similarly, irradiation of the derivatives **4b–d** (Scheme 9) gave the corresponding benzocycloheptenes **24b–d** along with **84b–c** (Structures 12); the latter was formed by a radical mechanism involving homolysis of the C-halogen bond followed by transannular hydrogen abstraction from one of the bridge

83

24a $R^1=R^2=H$

24b $R^1=R^2=Cl$

24c $R^1=Cl$, $R^2=Br$

24d $R^1=R^2=Br$

24e $R^1=H$, $R^2=Cl$

24f $R^1=H$, $R^2=Br$

84a R=H

84b R=Cl

84c R=Br

Structures 12

methylenes. In contrast to the Dewar type isomers of strained paracyclophanes, photolysis of **35**, the Dewar type isomer of [4]metacyclophane (**5**), did not afford **5** but the prismane type isomer **36** (Scheme 10) [33a].

As in the case of [6]paracyclophane (**1a**), irradiation of [6](1, 4)naphthalenophane (**55a**) produced the Dewar type isomer **53a** (s. Scheme 13) [40a]. In remarkable contrast, irradiation ($\lambda > 370$ nm) of anthracenophane **56a** did not give the Dewar type isomer **54a** (s. Scheme 13) but yielded five stereoisomers of cyclobutane dimers **85** (Structure 13) [67]. Four minor dimers have *cis-cis* junction around the cyclobutane ring, while the major product has *cis-trans* stereochemistry. This remarkable photochemical behavior of **56a** is ascribed to the high double bond character of the bond *a* in the anthracenophane system which is predicted by the semi-empirical molecular orbital calculations.

85

Structure 13

4.2 Electrophilic and Oxidative Reactions

[6]Paracyclophane (**1a**) undergoes acid-catalyzed rearrangement to the *meta* isomer **66** (Structures 6) which then isomerized to the *ortho* isomer, benzocyclooctene [5b, 10]. Recently, however, we found that considerable amount of the dimer **86** (Structures 14) was formed when the reaction was carried out at high concentration (10^{-1} M) [50, 68]. Formation of the dimer **86** was explained in terms of nucleophilic attack of **1a** to the carbocation intermediate before migration of the bridge takes place. Acid-catalyzed reaction of cyclophene **9a** (Structures 2) in methanol gave the *meta* isomer **73** (Structures 7) and methyl ether **87** (Structures 14) [28c]. The latter was produced by capture of the benzenonium ion intermediate by methanol. Similarly, treatment of photochemically generated **2a** (Structures 1) with TFA in THF afforded benzocycloheptene **24a** (Structures 12) and the labile 1,4-adduct **88a** (Structures 14) [69]. In methanol, adduct **88b** was obtained as the major product. In addition, photolysis of the Dewar type isomer **25a** of [4]paracyclophane (**3a**) in the presence of TFA in THF or methanol afforded the 1,4-adducts **26a** or **26d** (Scheme 8) [25]. The isomerization product, tetralin, was not observed. These gradual shift in the

reaction pattern was interpreted in terms of charge densities of the benzenonium ions and strain energies and proton affinities of the [n]cyclophanes [69].

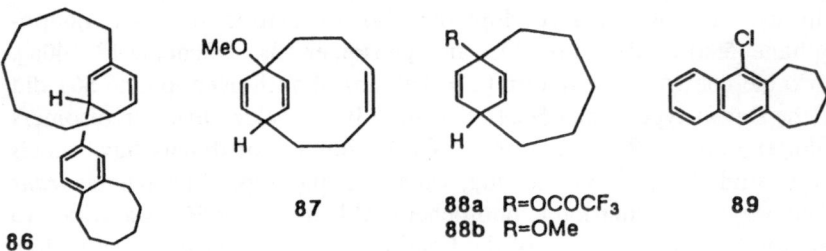

86   **87**   **88a** R=OCOCF$_3$ **89**
               88b R=OMe

Structures 14

[5]Metacyclophane (**4a**) behaves similarly; on treatment with TFA, **4a** and its derivatives **4b–d** (s. Scheme 9) gave the corresponding benzocycloheptenes **24a–d** (Structures 12) [66]. Monohalogenated products **24e–f** (Structures 12) were also obtained in the case of **4b–d**, which were formed by abstraction of a halonium ion from cationic intermediates. Similar isomerization of the naphthalene analogue **64** (s. Scheme 16) was effected by acetic acid, giving orthocyclophane **89** (Structures 14) [43].

In contrast to these, naphthalenophane **55a** and anthracenophane **56a** (s. Scheme 13) exhibited unusual reactivity in the acid-catalyzed reaction [50, 68]. Thus treatment of **55a** (10^{-3} M) with TFA gave dimers **76** and **90** as the major products. When the reaction was undertaken at high concentration (10^{-1} M), substantial amount of two stereoisomers (2:1 ratio) of the trimers **91** were obtained (Scheme 19). Moreover, similar reaction of **56a** (s. Scheme 13) gave only trimers **92** (Structure 15) (ca. 8:1 ratio) even when the reaction was undertaken under dilute conditions (10^{-3} M). These anomalous reactivity of **55a** and **56a** was explained in terms of (i) the relative stabilities of the carbocation intermediates and (ii) the affinities of **55a** and **56a** toward nucleophilic attack by the cationic intermediates.

With regard to the acid-catalyzed isomerization, 9,10-bridged anthracenophanes are capable of isomerizing to methylenedihydroanthracene (isotoluene) type isomers, since the energies of the tautomers would become more favorable as the size of the bridge decreases. Indeed, our attempt to synthesize [6]anthracenophane **60a** resulted in the formation of the tautomer **61a** presumably by isomerization of **60a** during reaction or isolation (Scheme 15) [42]. Although the tetramethyl derivative **60b** was isolated, it was very sensitive to acid; even trace amount induced clean isomerization to **61b**. In thermodynamic sense, severe steric repulsion around the methyl groups in **60b**, however, makes it more favorable to isomerize to **61b** than the case of **60a**.

Addition of bromine to [6]paracyclophane (**1a**) took place in the 1,4-fashion to give unstable adduct **93a** (Structures 16) [5b]. Similarly 1,4-adducts **93b–c** (Structures 16) were obtained by reaction of **1j** (Structures 2) with bromine and

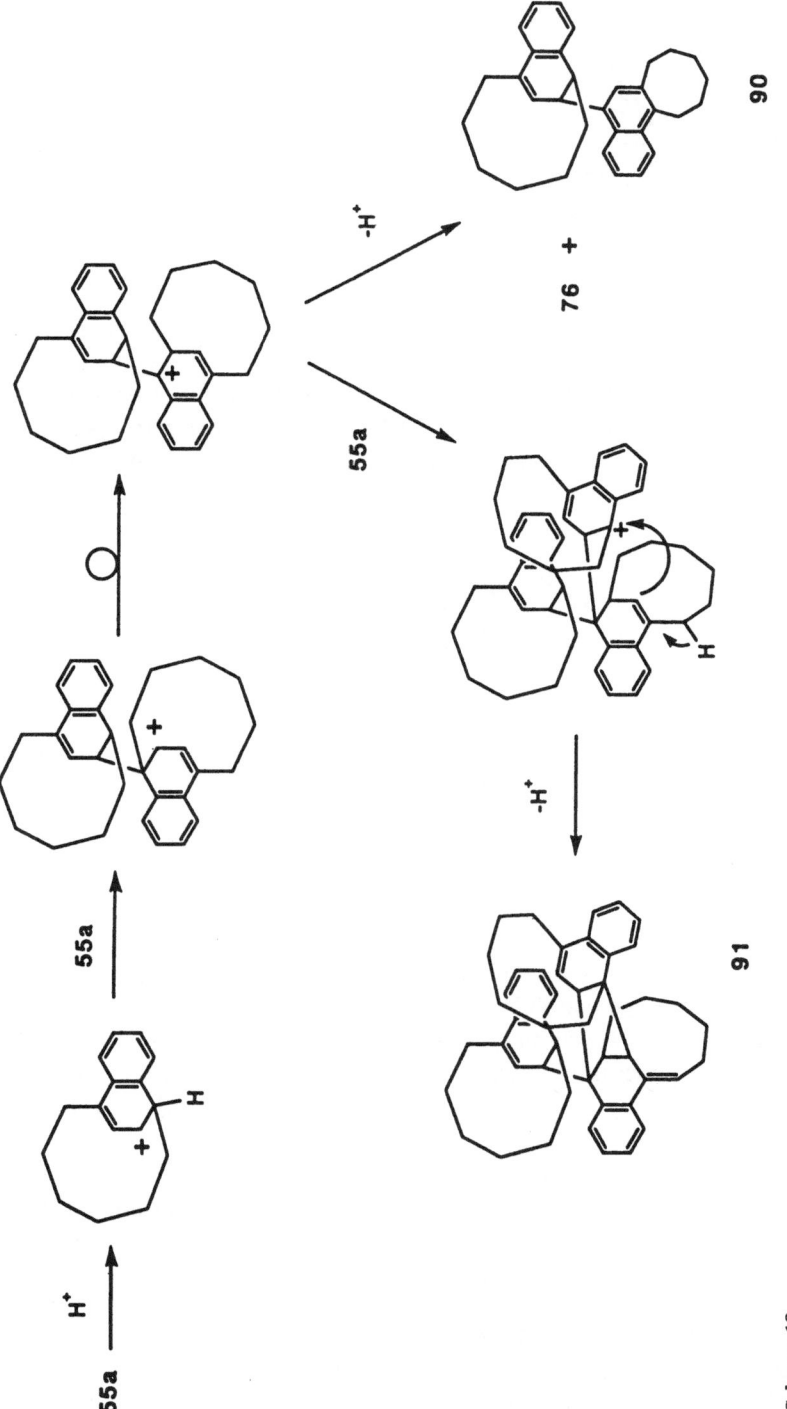

Scheme 19.

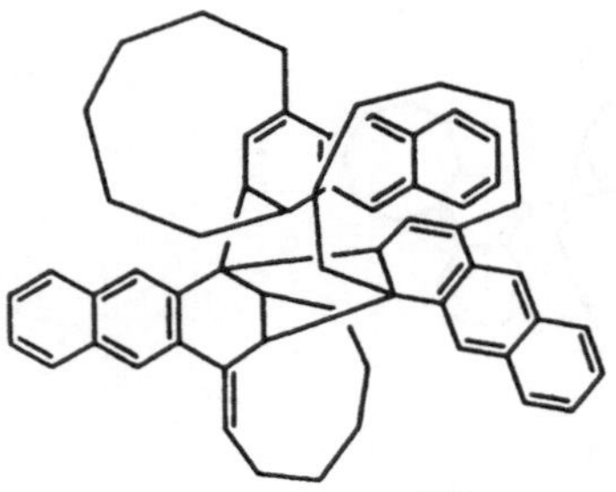

Yoshito Tobe

92

Structure 15

fuming nitric acid, respectively [13b, 64]. Whereas similar reaction of diester **1g** (Structures 2) afforded 1,2-adduct **94a** (Structures 16) together with 1,4-adduct **93c** in a ratio of 3:2. 1,2-Addition of osmium tetraoxide to **1g** was also reported to give diol **94b**.

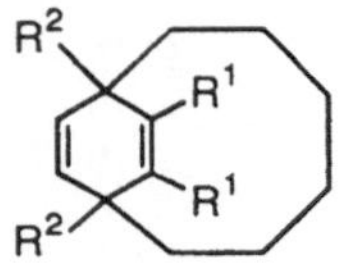

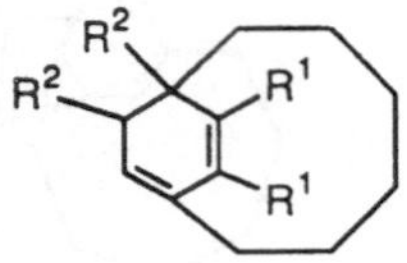

93a R^1=H, R^2=Br   94a R^1=CO$_2$Me, R^2=Br
93b R^1=CH$_2$OAc, R^2=Br   94b R^1=CO$_2$Me, R^2=OH
93c R^1=CH$_2$OAc, R^2=NO$_2$
93d R^1=CO$_2$Me, R^2=Br

Structures 16

Although it has been known that Diels-Alder reactions of [7]- and [8]para-cyclophanes with strong dienophiles such as DCA and hexafluoro-2-butyne are possible [28a, b], [6]paracyclophane (**1a**) and its derivatives undergo [4 + 2] addition with dienophiles more readily. Thus **1a** reacted with TAD, TCNE, and DCA to give the [4 + 2] adducts **95a–c** (Structures 17), respectively [5b, 50]. Similarly, its derivative **1g** (Structures 2) afforded similar adducts **95d–f** (Structures 17) with TAD, TCNE, and cyclooctyne, respectively [13b]. It should be pointed out that [8]paracyclophane did not give the Diels-Alder adduct with TCNE, but forms only a change-transfer complex [28b]. The enhanced reactivity of the [6]paracyclophane system is due to its high HOMO energy and deformed π bonds. A notable exception is the reaction of cyclophene **9a** (Structures 2) with TCNE which yielded the [2 + 2] adduct **96** (Structures 17) instead of a [4 + 2] adduct [28c]. Similarly, [4]paracyclophanes **29a–b** reacted with cyclopentadiene to give 2:1 adducts **30a–b**. In this reaction, **29a–b** served as a 2π component of the Diels-Alder reaction (Scheme 8) [26].

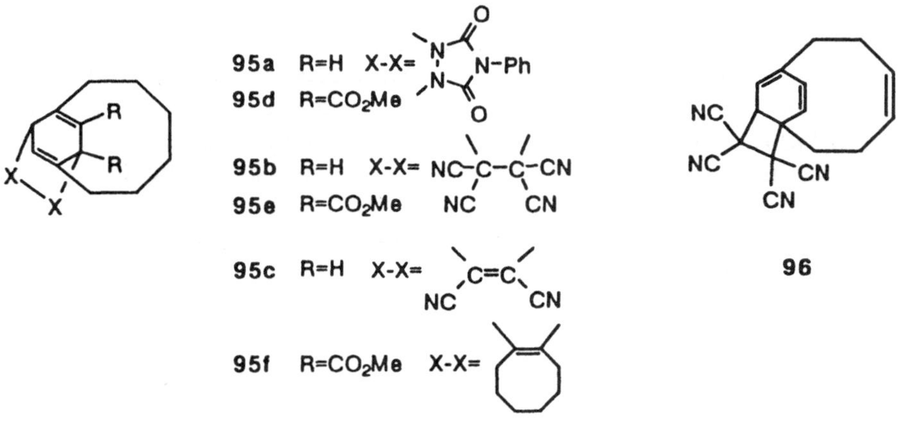

95a R=H   X-X= (N-N ring with N-Ph, O)

95d R=CO₂Me

95b R=H   X-X= NC-C-C-CN

95e R=CO₂Me   NC   CN

95c R=H   X-X= (C=C with NC, CN)

96

95f R=CO₂Me X-X= (cyclooctene ring)

Structures 17

[5]Metacyclophane (**4a**) also undergoes facile Diels-Alder reaction with weak dienophiles which did not react with **1a** [66, 70]. Thus reaction of **4a** with TCNE, DMAD, and MA furnished the [4 + 2] adducts **97a–c** (Structures 18), respectively. Similarly, halogen substituted derivatives **4b–d** (s. Scheme 9) gave the respective adducts **97d–j** (Structures 18), though these were less reactive than the parent hydrocarbon **4a**. This is ascribed to the electron-withdrawing effect of halogen substituents which lowers the HOMO level.

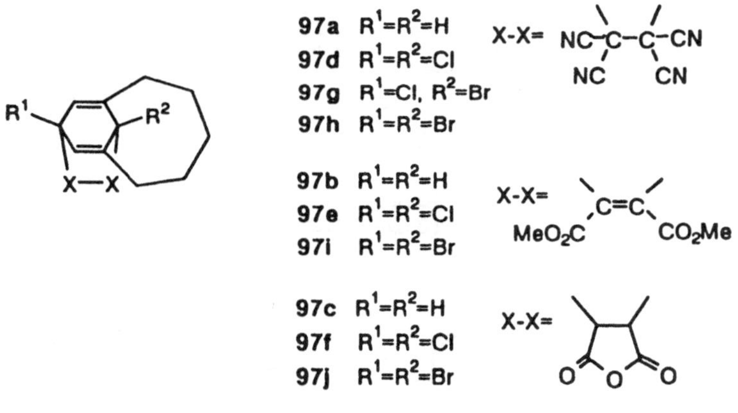

97a R¹=R²=H   X-X= NC-C-C-CN

97d R¹=R²=Cl   NC   CN

97g R¹=Cl, R²=Br

97h R¹=R²=Br

97b R¹=R²=H   X-X= (C=C with MeO₂C, CO₂Me)

97e R¹=R²=Cl

97i R¹=R²=Br

97c R¹=R²=H   X-X= (anhydride ring, O O O)

97f R¹=R²=Cl

97j R¹=R²=Br

Structures 18

Again, in contrast to the benzenophanes, the condensed benzenoid analogues **55a** and **56a** (s. Scheme 13) exhibited anomalous reactivity toward dienophiles [50, 71]. A remarkable effect of solvent on the reaction mode was also observed. Namely, reaction of **55a** and **56a** with TCNE gave [2 + 2] adducts **98a** and **99a** (Structures 19), respectively. Reaction with DCNA in dichloromethane also gave [2 + 2] adducts **98b** and **99b** (Structures 19). In benzene, ene-like products **100** and **101a** (Structures 19) were also obtained. Naphthalenophane **55a** reacted with DMAD in dichloromethane to give lactone **102** (Structures 19), while anthracenophane **56a** afforded lactone **103** and 2:1 adduct **104** (Structures 19). In benzene, **56a** gave ene-like product **101b** along with **103** and **104**. The formation of these unusual products was interpreted in terms of a mechanism involving zwitterion intermediates. Again, these anomalous reactivity reinforces the notion that this is due to high π bond order of the bond a and to their high HOMO energies.

Structures 19

Addition of 2 mole of dichlorocarbene to the [6]paracyclophane derivative **1g** (Structures 2) gave the cyclooctatetraenophane **105** as the major product together with the [8]paracycloph-1-ene **106** (Structures 20) [72]. Sequential addition of the carbene, first to **1g** to give a cycloheptatriene followed by the second one to this intermediate, accounts for the formation of the products.

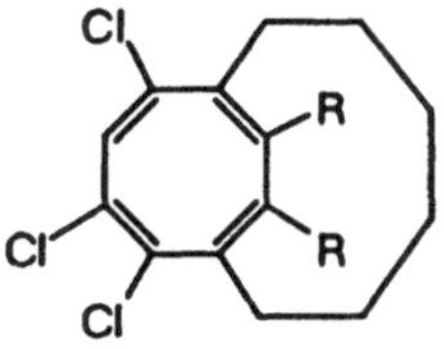

105 R=CO₂Me 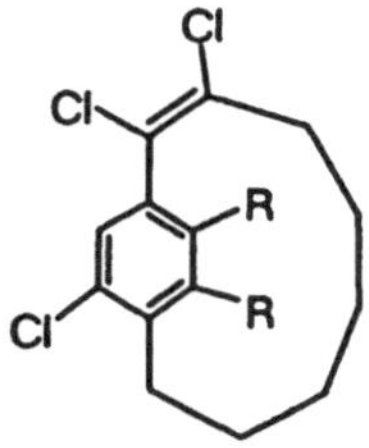   106 R=CO₂Me

Structures 20

[6]Paracyclophane (**1a**) suffers from peracid oxidation readily; oxidation of **1a** with MCPBA proceeded at 0 °C to yield the dimer **109** quantitatively [5b]. The formation of dimer **109** is explained in terms of [4 + 2] dimerization of the cyclohexadienone **108** which was formed by epoxide-carbonyl rearrangement of the initially formed epoxide **107** (Scheme 20). Naphthalenophane **55a** and anthracenophane **56a** (s. Scheme 13) were more reactive; the MCPBA oxidation completed immediately at − 78 °C to give unstable dienones **110** and **111** (Structures 21) [50].

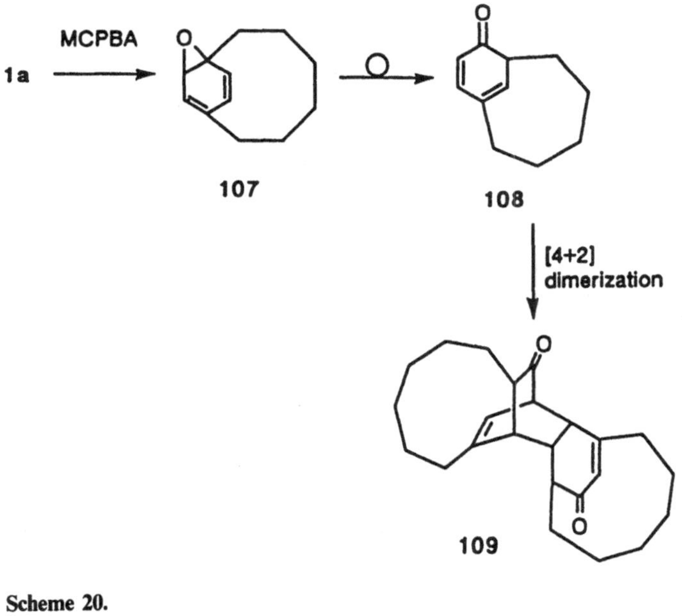

Scheme 20.

110   **111**

Structures 21

The central ring of bridged anthracene should be even more reactive; indeed, air oxidation of dithia[8]anthracenophane **42b** (Structures 4), whose aromatic ring is only slightly deformed from planarity, was observed giving anthraquinone presumably through an endoperoxide of **42b** [37].

4.3 Nucleophilic and Reductive Reactions

In comparison with the electrophilic reactions of strained [*n*]cyclophanes, much less works have been done on their reactivity toward nucleophiles. In view of the low LUMO energies and the strain release factor, it may well be expected that they show enhanced reactivity in nucleophilic reactions as well.

We investigated reactions of [6]paracyclophane (**1a**) with alkyllithium reagents and found out some unusual reactivities [73]. When **1a** was treated with *n*-BuLi in the presence of TMEDA in hexane at room temperature, 1,2- and 1,4-addition products **112a** and **113a–b** (Scheme 21) were obtained after quenching with H_2O or TMSCl. Similarly, treatment of **1a** with *s*-BuLi furnished adducts **112b** and **113c–d** (Scheme 21). These results indicated that **1a** suffers from nucleophilic addition of *n*-BuLi and *s*-BuLi, which represent the first example of such reaction of an unactivated benzene (without electron-withdrawing groups). In other words, **1a** is activated only by strain. On the other hands, treatment of **1a** with *t*-BuLi gave the metalation product **1p** (Structures 22) after quenching with TMSCl. However, when a large excess of *t*-BuLi was allowed to react for a long period of time, a formal substitution product **114a** was obtained (Scheme 21). Quenching with CO_2 followed by treatment with diazomethane yielded diester **114b** as the major product. This indicates that substitution takes place on lithiated species **1o** but not directly on **1a** (Scheme 21).

The finding that lithiation of **1a** took place selectively at the aromatic ring rather than the benzyl position opened a way for the functionalization of **1a** in a regioselective manner by means of metalation [58]. Thus **1a** was treated with 3 equiv. of the Lochmann's base (*t*-BuOK, *n*-BuLi) in hexane and then the reaction was quenched with appropriate electrophiles, giving monosubstituted products **1b**, **1l**, **1p**, and **1q** (Structures 2 and 22, Scheme 14) in good yields. When more excess base was employed, *para*-disubstitution products **1c**, **1m**, and **1r**

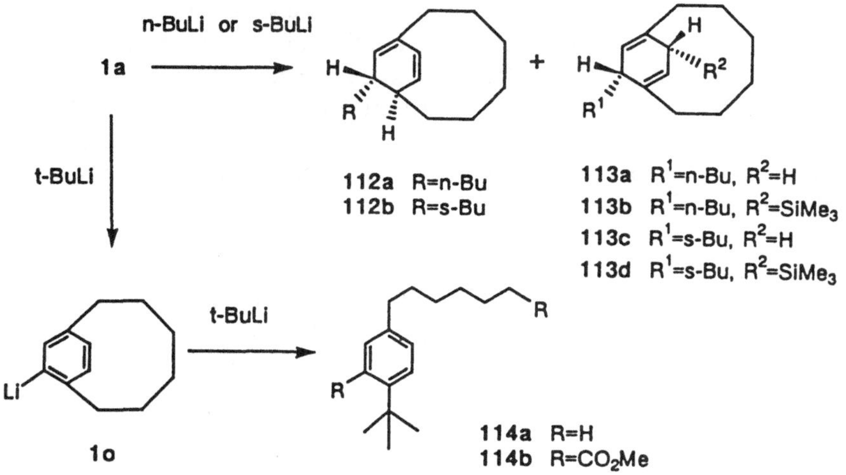

Scheme 21.

(Structures 2 and 22, Scheme 15) were obtained. The bromides **1l** and **1m** thus obtained were used for the further manipulation into the condensed benzenoid analogues (Schemes 14 and 15) [42]. However, when monobromide **1l** was treated with *n*-BuLi in THF followed by quenching with D_2O or methanol-*d*, no deuterium incorporation was observed in the product **1a** [41]. This means that single electron transfer rather than halogen-metal exchange takes place.

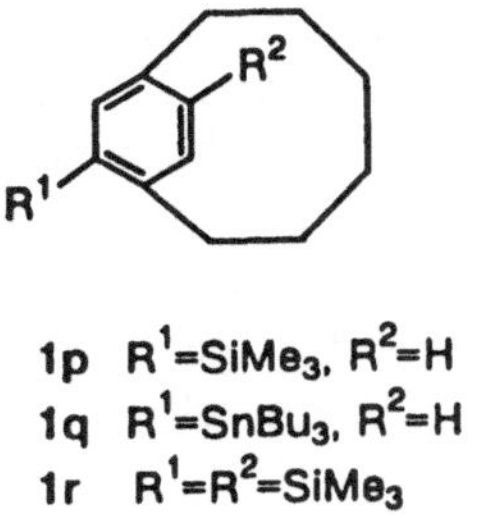

1p $R^1=SiMe_3$, $R^2=H$
1q $R^1=SnBu_3$, $R^2=H$
1r $R^1=R^2=SiMe_3$

Structures 22

Even such weak nucleophile as alcohols underwent nucleophilic addition to highly strained [4]paracyclophane (**3a**), which was generated by photochemical valence isomerization of the Dewar isomer **25a** (Scheme 8) [24].

Although nucleophilic reactions of the smallest [*n*]metacyclophane **4a** (Structures 1) was not reported, unusual reactivity of its halogen-substituted derivatives **4b–e** (Scheme 9, Structures 23) toward nucleophiles was reported by Bickelhaupt [66, 74]. Treatment of chlorocyclophanes **4b** and **4e** with sodium alkoxide gave nucleophilic substitution products **4f** and **4g** (Structures 23), respectively, through a S_NAr mechanism [74b]. The regioselective substitution at C-11 is explained in terms of (i) greater stability of the Meisenheimer complex than that derived from C-8 attack, (ii) larger LUMO coefficient at this position, and (iii) greater pyramidalization of C-11 than that of C-8. On the other hand, reaction of **4b** and **4e** with NaOH resulted in the formation of rearranged products **115a** and **115b** (Structures 23), respectively. The formation of **115a–b** is interpreted by the initial attack of the hydroxyl anion at the bridgehead position followed by migration of the bridge since similar reaction was observed with the bromohydrin **116** (Structures 23) [75].

4e R^1=H, R^2=Cl	**115a** R=H	**116**
4f R^1=H, R^2=OR	**115b** R=Cl	
4g R^1=Cl, R^2=OR		
4h R^1=Cl, R^2=t-Bu		

Structures 23

Reactivity of **4b–d** (s. Scheme 9) toward alkyllithiums varies substantially depending on the substituents [66, 74]. Thus, while **4b** did not react with *n*-BuLi, **4c** gave tricyclic compound **84b** (Structures 12) through a single electron transfer mechanism. This was confirmed from the fact that reduction of **4b–d** with NaH/Ni(OAc)$_2$ gave tricyclic hydrocarbon **84a**. On the other hand, reaction of **4b** with *t*-BuLi afforded formal substitution product **4h** (Structures 23) like the reactions with RONa. Halogen-metal exchange, however, was the major reaction pathway of dibromide **4d** with *t*-BuLi. Bromochloro derivative **4c** gave tricyclic compound **84b** through a single electron transfer. Low LUMO energies and a large amount of strain released in the initial stage of the reactions are responsible for the enhanced reactivity of these [5]metacyclophane derivatives.

Catalytic hydrogenation of [6]paracyclophane **1g** (Structures 2) afforded bridgehead alkene **117** (Structures 24), by uptake of 2 mole of hydrogen [13b]. Since hyperstable bridgehead alkenes are known to resist further hydrogenation, the reaction ceased at this stage. Attempted dehalogenation of 8, 11-

117   **118**

Structures 24

dichloro[5]metacyclophane (**4b**) under catalytic reduction conditions, however, resulted in the formation of completely hydrogenated product **118** (Structures 24) [66].

5 Concluding Remarks

During the last five years, enormous achievements have been made in the chemistry of strained [*n*]cyclophanes owing to the development of new synthetic strategies. Species that had been thought as "impossible" were synthesized or characterized spectroscopically. Through the thorough experimental investigations of their spectroscopic properties, structure, and reactivities combined with sophisticated theoretical studies, now it becomes possible to draw the following conclusions: (i) The strained [*n*]cyclophanes have strong tendency to maintain their aromatic character even though the geometrical strain increases to keep the favorable geometry to retain aromaticity. (ii) The unusual reactivities most of which seems as if they were due to the "cyclohexatriene" character should be ascribed to increased strain because of the geometry deformation but not to localization of the π bonds.

More strained molecules with a smaller bridge are the next obvious targets, but also are more constrained systems such as [1.1]metacyclophane (**119**) [76], [1.1]paracyclophane (**120**) [77], and adamantanophane (**121**) (Structures 25) [78]. Because of structural variety that allows one to modify the mode of deformation, distorted condensed benzenoid (and non-benzenoid as well) aromatics will continue to be a source of exotic reactions. These molecules will continue to accept challenge of synthetic chemists because new alternative methods should always be developed. Moreover, these investigations of special π bond systems will certainly deepen our understanding of chemical bonding, but also would provide opportunities of finding useful reactions and functions of organic molecules.

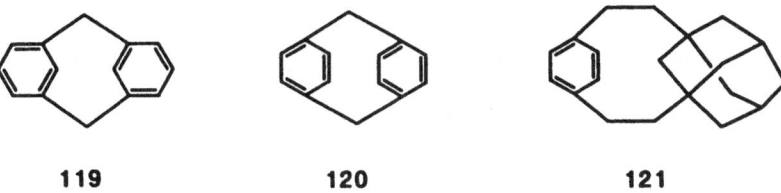

119   **120**   **121**

Structures 25

Acknowledgement: The author is grateful to the collaborators in his research group whose names are found in the references. Without their contribution these accomplishments would not have been achieved. Thanks are also due to Professors F. Bickelhaupt, R. Gleiter, M. Jones, Jr., Y. Kai, N. Kasai, S. Sternhell, and W. de Wolf for their pleasant and fruitful cooperation.

6 References

1. Kane VV, Wolf AD, Jones Jr. M (1974) J. Am. Chem. Soc. 96: 2643
2. Cram DJ, Cram JM (1971) Acc. Chem. Res. 4: 204
3. For reviews; Greenberg A, Liebman JF (1978) Strained organic molecules. Academic Press, New York, p 153; (b) Rosenfeld SM, Choe KA (1983) In: Keehn PM, Rosenfeld SM (eds) Cyclophanes, Academic Press, New York, vol 1, p 311; (c) Bickelhaupt F, de Wolf WH (1988) Recl. Trav. Chim. Pays-Bas 107: 459; (d) Bickelhaupt, F (1990) Pure Appl. Chem. 62: 373; (e) Bickelhaupt F, de Wolf WH (1993) In: Halton B (ed) Advances in strain in organic chemistry. JAI Press, Greenwich, Connecticut, vol 3, p 185; (f) Kane VV, de Wolf WH, Bickelhaupt F, Tetrahedron (to be published)
4. Kammula SL, Iroff LD, Jones Jr. M, van Straten JW, de Wolf WH, Bickelhaupt F (1977) J. Am. Chem. Soc. 99: 5815
5. (a) Tobe Y, Kakiuchi K, Odaira Y, Hosaki T, Kai Y, Kasai N (1983) J. Am. Chem. Soc. 105: 1376; (b) Tobe Y, Ueda K-I, Kakiuchi K, Odaira Y, Kai Y, Kasai N (1986) Tetrahedron 42: 1851
6. Tobe Y, Nakayama A, Kakiuchi K, Odaira Y, Kai Y, Kasai N (1987) J. Org. Chem. 52: 2639
7. Tobe Y, Furukawa T, Kobiro K, Kakiuchi K, Odaira Y (1989) J. Org. Chem. 54: 488
8. (a) Tobe Y, Ueda K-I, Odaira Y (1986) Angew, Chem., Int. Ed. Engl. 25: 369; (b) Tobe Y, Kaneda T, Kakiuchi K, Odaira Y (1986) Chem. Lett. 1217; (c) Tobe Y, Ueda K-I, Kaneda T, Kakiuchi K, Odaira Y, Kai Y, Kasai N (1987) J. Am. Chem. Soc. 109: 1136
9. Gleiter R, Treptow B (1990) Angew. Chem., Int. Ed. Engl. 29: 1427
10. Tobe Y, Ueda K-I, Kakiuchi K, Odaira Y (1983) Chem. Lett. 1645
11. Sakai Y, Toyotani S, Tobe Y, Odaira Y (1979) Tetrahedron Lett. 3855; Sakai Y, Tobe Y, Odaira Y (1980) Chem. Lett. 691; Sakai Y, Toyotani S, Ohtani M, Matsumoto M, Tobe Y, Odaira Y (1981) Bull. Chem. Soc. Jpn. 54: 1474
12. Tobe Y, Ueda Y, Matsumoto M, Sakai Y, Odaira Y (1982) Tetrahedron Lett. 23: 537; Tobe Y, Kishimura T, Kakiuchi K, Odaira Y (1983) J. Org. Chem. 48: 551
13. (a) Liebe J, Wolff C, Tochtermann W (1982) Tetrahedron Lett. 23: 171; (b) Liebe J, Wolff C, Krieger C, Weiss J, Tochtermann W (1985) Chem. Ber. 118: 4144
14. Jessen JL, Wolff C, Tochtermann W (1986) Chem. Ber. 119: 297
15. Beitz G, Vagt U, Tochtermann W (1985) Tetrahedron Lett. 26: 721; Tochtermann W, Vagt U, Snatzke G (1985) Chem. Ber. 118: 1996; Tochtermann W, Olsson G, Vogt C, Peters E-M, Peters K, von Schnering HG (1987) Chem. Ber. 120: 1523
16. Bockish F, Dreeskemp H, von Haugwitz T, Tochtermann W (1991) Chem. Ber. 124: 1831
17. Miki S, Shimizu R, Nakatsuji H (1992) Tetrahedron Lett. 33: 953
18. Nitta M, Kobayashi T (1984) Tetrahedron Lett. 25: 959
19. van Straten JW, Landheer IJ, de Wolf WH, Bickelhaupt F (1975) Tetrahedron Lett. 4499; van Straten JW, Turkenburg LAM, de Wolf WH, Bickelhaupt F (1985) Recl. Trav. Chim. Pays-Bas 104: 89
20. Tobe Y, Kakiuchi K, Kobiro K, Odaira Y (1993) Synthesis 460
21. Jessen JL, Schröder G, Tochtermann W (1985) Chem. Ber. 118: 3287; Jenneskens LW, Kostermans GBM, ten Brink HJ, de Wolf WH, Bickelhaupt F (1985) J. Chem. Soc., Perkin Trans. I 2119
22. Jenneskens LW, de Kanter FJJ, Kraakman PA, Turkenburg LAM, Koolhaas WE, de Wolf WH, Bickelhaupt F, Tobe Y, Kakiuchi K, Odaira Y (1985) J. Am. Chem. Soc. 107: 3716
23. Tobe Y, Kaneda T, Kakiuchi K, Odaira Y (1985) Chem. Lett. 1301; Kostermans GBM, de Wolf WH, Bickelhaupt F (1986) Tetrahedron Lett. 27: 1095; Kostermans GBM, de Wolf WH, Bickelhaupt F (1987) Tetrahedron 43: 2955
24. Tsuji T, Nishida S (1987) J. Chem. Soc., Chem. Commun. 1189; Tsuji T, Nishida S (1988) J. Am. Chem. Soc. 110: 2157

25. Kostermans GBM, Bobeldijk M, de Wolf WH, Bickelhaupt F (1987) J. Am. Chem. Soc. 109: 2471
26. Tsuji T, Nishida S (1989) J. Am. Chem. Soc. 111: 368
27. Fujita S, Hirano S, Nozaki H (1972) Tetrahedron Lett. 403; Hirano S, Hiyama T, Fujita S, Nozaki H (1972) Chem. Lett. 707; Hirano S, Hara H, Hiyama T, Fujita S, Nozaki H (1975) Tetrahedron 31: 2219
28. (a) Noble K-L, Hopf H, Jones Jr. M, Kammula SL (1978) Angew. Chem., Int. Ed. Engl. 17: 602; (b) Noble K-L, Hopf H, Ernst L (1984) Chem. Ber. 117: 455; (c) Tobe Y, Sorori T, Kobiro K, Kakiuchi K, Odaira Y (1987) Tetrahedron Lett. 28: 2861
29. Shea KL, Burke LD, Doedens RJ (1985) J. Am. Chem. Soc. 107: 5305
30. Goodman JL, Berson JA (1985) J. Am. Chem. Soc. 107: 5424
31. (a) van Straten JW, de Wolf WH, Bickelhaupt F (1977) Tetrahedron Lett. 4667; (b) Turkenburg LAM, Blok PML, de Wolf WH, Bickelhaupt F (1981) Tetrahedron Lett. 22: 3317; (c) Turkenburg LAM, de Wolf WH, Bickelhaupt F (1983) Tetrahedron Lett. 24: 1817; (d) Jenneskens LW, de Kanter FJJ, Turkenburg LAM, de Bore HJR, de Wolf WH, Bickelhaupt F (1984) Tetrahedron 40: 4401
32. Turkenburg LAM, van Straten JW, de Wolf WH, Bickelhaupt F (1980) J. Am. Chem. Soc. 102: 3256
33. (a) Kostermans GBM, Hogenbirk M, Turkenburg LAM, de Wolf WH, Bickelhaupt F (1987) J. Am. Chem. Soc. 109: 2855; (b) Kostermans, GBM, van Dansik P, de Wolf WH, Bickelhaupt F (1987) J. Am. Chem. Soc. 109: 7887; (c) Kostermans GBM, van Dansik P, de Wolf WH, Bickelhaupt F (1988) J. Org. Chem. 53: 4531
34. Wiberg KB, O'Donnell MJ (1979) J. Am. Chem. Soc. 101: 6660
35. Haley Jr. JF, Keehn PM (1973) Tetrahedron Lett. 4017
36. Rosenfeld SM, Sanford EM (1987) Tetrahedron Lett. 28: 4775
37. Chung J, Rosenfeld SM (1983) J. Org. Chem. 48: 387
38. Ohta M, Tamura S, Okajima T, Fukazawa Y (1992) Abstract of 63th Annual Meeting of the Chemical Society of Japan, 3B234
39. Hunger J, Wolff C, Tochtermann W, Peters E-M, Peters K, von Schering HG (1986) Chem. Ber. 119: 2698
40. (a) Tobe Y, Takahashi T, Kobiro K, Kakiuchi K (1990) Chem. Lett. 1587; (b) Tobe Y, Takahashi T, Ishikawa T, Yoshimura M, Suwa M, Kobiro K, Kakiuchi K, Gleiter R (1990) J. Am. Chem. Soc. 112: 8889
41. Tobe Y (unpublished results)
42. Tobe Y, Ishii H, Saiki S, Kakiuchi K, Naemura K (1993) J. Am. Chem. Soc. 115: 11604
43. Grice P, Reese CB (1980) J. Chem. Soc., Chem. Commun. 424
44. (a) Remington RB, Lee TJ, Schaefer III HF (1986) Chem. Phys. Lett. 124: 199; (b) Rice JE, Lee TJ, Remington RB, Allen WD, Clabo DA, Schaefer III HF (1987) J. Am. Chem. Soc. 109: 2902; (c) Lee TJ, Rice JE, Allen WD, Remington RB, Schaefer III HF (1988) Chem. Phys. 123: 1
45. Jenneskens LW, Lowen JN, de Wolf WH, Bickelhaupt F (1990) J. Phys. Org. Chem. 3: 295
46. Grimme S (1992) J. Am. Chem. Soc. 114: 10542
47. von Armin M, Peyerimhoff SD (1993) Thor. Chim. Acta 85: 43
48. Wong C-K, Cheung Y-S, Wong HMC, Li W-K (1993) J. Chem. Res. (S) 308
49. (a) Jenneskens LW, de Kanter FJJ, de Wolf WH, Bickelhaupt F (1987) J. Comput. Chem. 8: 1154; (b) Jenneskens LW, Louwen JN, Bickelhaupt F (1989) J. Chem. Soc., Perkin Trans. II 1893
50. Tobe Y, Takemura A, Jimbo M, Takahashi T, Kobiro K, Kakiuchi K (1992) J. Am. Chem. Soc. 114: 3479
51. Hart H, Nwokogu G (1981) J. Org. Chem. 46: 1251
52. Günther H, Schmitt P, Fischer H, Tochtermann W, Liebe J, Wolff C (1985) Helv. Chim. Acta 68: 801
53. Wolff C, Liebe J, Tochtermann W (1982) Tetrahedron Lett. 23: 1143
54. Tobe Y, Kawaguchi M, Kakiuchi K, Naemura K (1993) J. Am. Chem. Soc. 115: 1173
55. Bodwell G, Ernst L, Haenel MW, Hopf H (1989) Angew. Chem., Int. Ed. Engl. 28: 455
56. Turkenburg LAM, de Wolf WH, Bickelhaupt F, Cofino WP, Lammertsma K (1983) Tetrahedron Lett. 24: 1821
57. Gready JE, Hambley TW, Kakiuchi K, Kobiro K, Sternhell S, Tansey CW, Tobe Y (1990) J. Am. Chem. Soc. 112: 7538
58. Tobe Y, Jimbo M, Ishii H, Saiki S, Kakiuchi K, Naemura K (1993) Tetrahedron Lett. 34: 4969

59. van Zijl PCM, Jenneskens LW, Bastiaan EW, MacLean C, de Wolf WH, Bickelhaupt F (1986) J. Am. Chem. Soc. 108: 1415
60. Jenneskens LW, Klamer JC, de Boer HJR, de Wolf WH, Bickelhaupt F, Stam CH (1984) Angew. Chem., Int. Ed. Engl. 23: 238
61. van Straten JW, Landheer IJ, de Wolf WH, Bickelhaupt F (1975) Tetrahedron Lett. 4499
62. Jenneskens LW, de Wolf WH, Bickelhaupt F (1986) Tetrahedron 42: 1571
63. Murray DF, Baum MW, Jones Jr. M (1986) J. Org. Chem. 51: 1
64. Liebe J, Wolff C, Tochtermann W (1982) Tetrahedron Lett. 23: 2439
65. Miki S, Abdel-Latif FM, Nakayama T, Hamanoue K (1993) Proc. Indian Acad. Sci. in press.
66. Jenneskens LW, de Boer HJR, de Wolf WH, Bickelhaupt F (1990) J. Am. Chem. Soc. 112: 8941
67. Tobe Y, Takahashi T, Kobiro K, Kakiuchi K (1991) J. Am. Chem. Soc. 113: 5804
68. Tobe Y, Jimbo M, Kobiro K, Kakiuchi K (1991) J. Org. Chem. 56: 5241
69. Kostermans GBM, Kwakman PJ, Pouwels PJW, Somsen G, de Wolf WH, Bickelhaupt F (1989) J. Phys. Org. Chem. 2: 331
70. Turkenburg LAM, Blok PML, de Wolf WH, Bickelhaupt F (1982) Angew. Chem., Int. Ed. Engl. 21: 298
71. Tobe Y, Takahashi T, Kobiro K, Kakiuchi K (1991) Tetrahedron Lett. 32: 359
72. Königstein V, Tochtermann W, Peters E-M, Peters K, von Schnering HG (1987) Tetrahedron Lett. 28: 3483
73. Tobe Y, Jimbo M, Saiki S, Kakiuchi K, Naemura K (1993) J. Org. Chem. 58: 5883
74. (a) Jenneskens LW, Klamer JC, de Wolf WH, Bickelhaupt F (1984) J. Chem. Soc., Chem. Commun. 733; (b) Kraakman PA, Valk J-M, Niederländer HAG, Brouwer DBE, Bickelhaupt FM, de Wolf WH, Bickelhaupt F, Stam CH (1990) J. Am. Chem. Soc. 112: 6638
75. Grice P, Reese CB (1979) Tetrahedron Lett. 2563
76. Wijsman GW, van Es DS, de Wolf WH, Bickelhaupt F (1993) Angew. Chem., Int. Ed. Enhgl. 32: 726
77. Tsuji T, Ohkita M, Nishida S (1993) J. Am. Chem. Soc. 115: 5284
78. Lemmerz R, Nieger M, Vögtle F (1993) J. Chem. Soc., Chem. Commun. 1168

Transition Metal Complexes of (Strained) Cyclophanes

J. Schulz, F. Vögtle

Institut für Organische Chemie und Biochemie der Universität Bonn,
Gerhard-Domagk Straße 1, 53121 Bonn, FRG

Table of Contents

Topics in Current Chemistry, Vol. 172
© Springer-Verlag Berlin Heidelberg 1994

Numerous transition metal complexes of strained cyclophanes are known, including the metals Cr, Mo, W, Fe, Ru, Co, Rh, Ir, Ni, Cu, Ag and U. The complexed phanes turned out to be valuable subjects for NMR-examinations in order to draw a map of the magnetic environment of sandwich complexes, e.g. di(benzene)chromium and ferrocene, and they provided an insight into the effect of ring current disrupture, due to metal complexation. First attempts were carried out to derivatize the cyclophanes using the metal fragment as an auxiliary group, which is easily introduced as well as removed. In this context a tricarbonylchromium complex of a chiral [2.2]metacyclophane was prepared recently and its absolute configuration, containing two elements of chirality, was determined by X-ray crystallography. Complexation of the cyclophanes was also used for the purpose of the preparation of numerous novel phanes which were not yet available via other routes, e.g. *syn*-[2.2]metacyclophane, and phanes, which contain *anti*-aromatic π-decks and are only stable when complexed with the metal. Removing the stabilizing metal unit lead to valence isomerism, which gave interesting hydrocarbons, e.g. propella[3₄]prismane.

The synthesis of [4]ferrocenophane, containing an iron-II-ion in the center of the hydrocarbon cage, has been achieved recently. The closely related [3]- and [5]ferrocenophanes remain unknown yet, but intramolecular linkage of the ferrocene nucleus with four of the five bridges has already been successfully carried out. The lengths of the bridges as well as their positions strongly influence the geometry of the ferrocenophanes. The electron spectra in the [4ₙ]ferrocenophane series interestingly show a linear relationship of the absorption towards shorter wave length with increase in the number of bridges.

Of increasing interest is the construction of an organometallic polymer, containing alternating phane and metal units, because of its use as an electrical conductor. The first oligomeric units of such a polymer have already been prepared, using either iron-II, coordinated to indenophanes, or ruthenium-II, coordinated sandwich-like to multibridged [2ₙ]cyclophanes. The 2-electron reduction product of the sandwich complex of the bis(ruthenium-II) complex with [2₄](1,2,4,5)cyclophane, the most stable in this series, establishes the first example, where a netto-2-electron intervalence transfer of a discrete, mixed-valence organometallic complex was observed.

1 Introduction, Scope

The chemistry of cyclophanes, the first representative of which, [2.2]meta-cyclophane was prepared by Pellegrin [1] as early as 1899, was first systematically studied in the 1950s, mainly by the innovative work of Cram [2]. Through specific syntheses, he prepared new members and recognized characteristic differences with regard to open chained benzene derivatives.

All members in the series of [2ₙ]cyclophanes [3] up to the sixfold-clamped superphane [4] are known today. Among the [n]paracyclophanes [5] bridging with a pentamethylene chain has established an end point so far and in the case of the syntheses of multilayered [2ₙ]cyclophanes [6] a limit of six stacked benzene rings has been achieved.

Of current importance in the chemistry of cyclophanes is their application as complex ligands for metals. Development of new agents which transfer transition metal moieties to arenes is progressing and the number of π-complexes of arenes with main group metals [7] is also increasing.

Complexation of cyclophanes with metals is important with regard to the effect of the change in ring current which accompanies this metal complexation, as well as the change of the chemical properties of the arene and the coordinated metal atom. A promising research field at present is the construction of a one dimensional metalorganic polymer consisting of a chain of alternating cyclophane- and metal-units. Metal coordination facilitates the syntheses of cyclophanes which cannot be prepared by other methods, and of those which are stable only when coordinated to the metal.

As early as 25 years ago, a comprehensive overview [8] of the chemistry of ferrocenophanes was published. In this particular field, development is so extensive that in a recent survey [9] only certain aspects were considered. In this progress report we will restrict ourselves to the complexation chemistry of cyclophanes with transition metals. Complexation chemistry with main group metals mainly consists of structural descriptions at present. As to complexes of cyclophanes with elements of the third [10], fourth [10a, 11] and fifth [12] main groups we refer to the references given.

2 Nomenclature

The family of *phanes* includes all cyclic compounds containing at least one "aromatic" nucleus (arene) and one aliphatic bridge [13, 14]. Phanes are divided into *carbophanes* (with a carbocyclic-) and *heterophanes* (with a heterocyclic-) aromatic nucleus. Phanes containing a benzene ring are called cyclo- (or benzeno-)phanes. A metallocenophane is built up of a metallocene [9]. The length of the bridge in a phane, which can be zero in case of a single bond only, is given in square brackets at the beginning of its name followed by the substitution pattern given in round brackets. The name of the phane is determined by the name of the bridged aromatic nucleus (arene). Cyclopentadienyl ligands can be bridged intraannularly as in 5 or interannularly as in 4 (Fig. 1) [13].

In the case of multibridged "arenes", or when two arene rings are linked with more bridges, the length of each bridge, followed by the substitution pattern, is indicated (Fig. 1). In the general formula $[m_n](u, v)$phane, m expresses the length of the bridge, n the number of bridges with the length m; u and v indicate the substitution pattern of the arene [9, 13, 14]. The nomenclature of metal complexes of (cyclo-)phanes in the literature is not uniform: In the case of a bridged metallocene the name of the phane is referred to the name of the parent metallocene, whereas all other bridged sandwich complexes are named after the parent phane, although analogous structural units are present (Fig. 2) [15–19].

It is not relevant in nomenclature whether the respective metal free ligand has a phane structure as in complexes 10 and 11, or not, as in *ansa*-compounds 12 [17] and 15 [19] (Fig. 2).

Complicated and less clear is the situation in complexes like 14 [18] (Fig. 2). It consists of two ferrocene units, both of which are bridged intraannularly and therefore this arrangement is designated as ferrocenophane [13, 18]. A different way of looking at it, considering it as the double decker sandwich complex of dipotassium $[3_2](1,3)$cyclopentadienylophanate (16), would not lead to an appropriate name (Fig. 2) [13]. Replacing two cyclopentadienyl rings with

J. Schulz and F. Vögtle

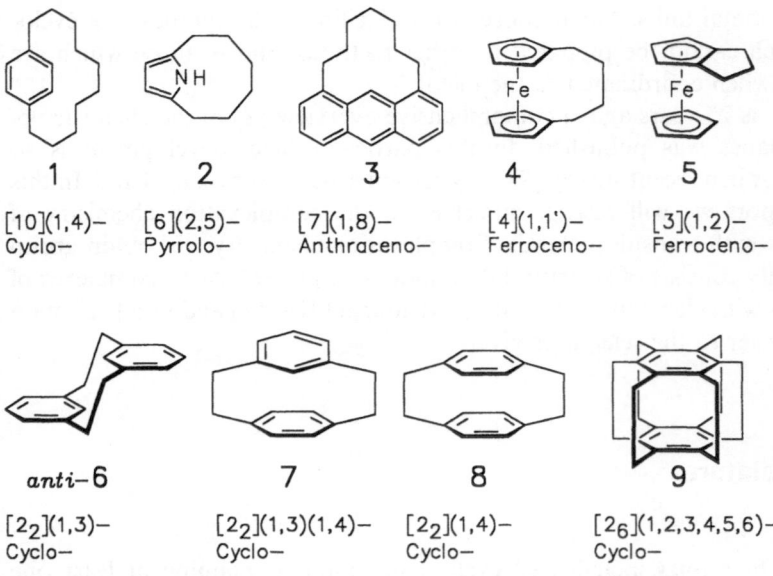

Fig. 1. Nomenclature of phanes (the syllable-phane is left out) [13, 14]

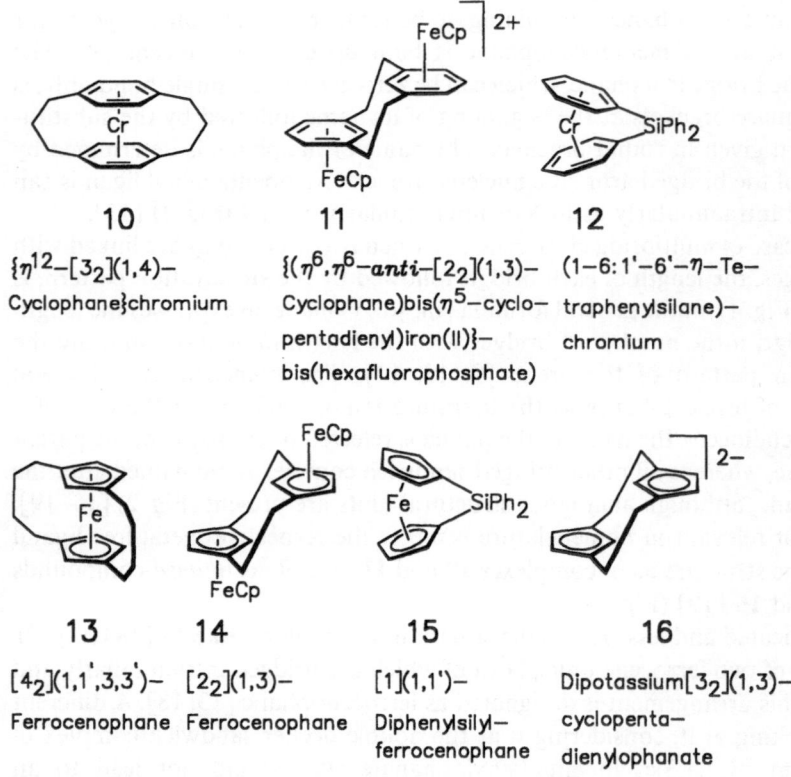

Fig. 2. Metallocyclophanes and metallocenophanes [15–19]

benzene rings results in the structurally related compound **11** (Fig. 2) [16], and now its name expresses that it is an iron sandwich complex of *anti-***6**. The term *"metallocyclophane"* includes all metal complexes of cyclophanes [20]; within this class of compounds ferrocenophanes hold a special position. In this report the syntheses and selected characteristic properties of transition metal complexes of strained (cyclo-)phanes by means of ligands having phane structure when uncomplexed, are given. The group of *ansa*-metallocenophanes [21] as well as crown ether linked ferrocenes [22], which establish connections to "supramolecular chemistry", will not be considered. Two fundamental types of complexes can be distinguished:

a) metal carbonyl- and related complexes
b) sandwich complexes with
   – interannularly bridged ligands and
   – intraannularly bridged ligands

3 Syntheses

Four possible general structural types are illustrated in Fig. 3. Structure types A and B are represented by the mono- and bis(tricarbonyl-chromium) complexes ($ML_x = Cr(CO)_3$) and double decker and triple decker sandwich complexes (e.g. $ML_x = FeCp^{1+}$; $Ru(C_6H_6)^{2+}$; $CoCp^{1+}$). The benzene rings of the cyclophane ligands can be replaced by cyclopentadienyl- or cyclobutadienyl-ligands.

Representatives of structure type C are also realized ($M = Cr^0$, Fe^{2+}); exchange of both benzene rings by cyclopentadienyl rings leads to the family of ferrocenophanes (fcp), of which a large number of representatives are known. Interesting electrochemical properties are to be expected with structure type D. The syntheses of several monomeric units, where a metal atom is coordinated to two cyclophane ligands ($M = Cr^0$, Fe^{2+}, Ru^{2+}), constitute a partial success on the way to a corresponding polymer ($n = \infty$).

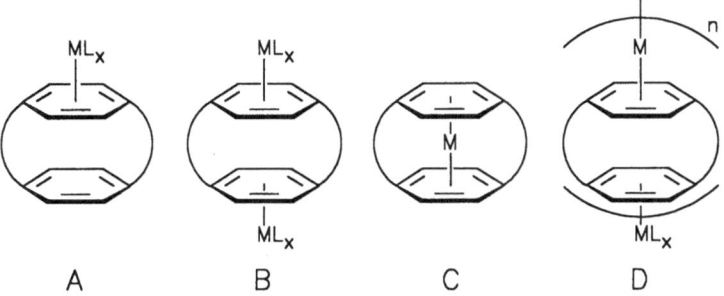

Fig. 3. General structural types of metallocyclophanes

3.1 Carbonyl- and Related Complexes

3.1.1 General Remarks

Group VIb elements provide by far the most representatives of carbonyl complexes of cyclophanes. Their preparation is achieved by heating the metal hexacarbonyl $M(CO)_6$ (M = Cr, Mo, W) with the respecting cyclophane in a high boiling point solvent, such as ligroin or di-n-butylether, for several hours [23]. Fischer and Öfele [24] prepared the parent compound, (η^6-benzene)-$Cr(CO)_3$, in this way. The reaction conditions are drastic; better yields are obtained in a mixture of di-n-butylether/THF [25].

By heating the metal hexacarbonyl in a solvent (L) with electron donor properties, only three carbonyl groups are replaced and the resulting $M(CO)_3L_3$ complexes (L = NH_3 [26a], acetonitrile [26b,c], pyridine [26d], L_3 = naphthalene [26e]) are much gentler complexing agents [23. 26]. Annelated arenes, like naphthalene, are weaker ligands than benzene and can be exchanged with the latter [27].

3.1.2 Carbonyl Complexes of Cyclophanes Containing Aromatic π-Decks

[n](1,4)Cyclophanes (*ansa*-phanes) are interesting ligands because the degree of their distorted benzene ring can be adjusted by the length of the bridge. The cyclophane with the shortest bridge known at present is [5](1,4)cyclophane (**17**) [5], which is however unstable above $-70\,°C$ and therefore unsuitable for complexation. Its next higher homolog, [6](1,4)cyclophane (**18**) [28b], was transferred to its $Cr(CO)_3$ complex **18a** recently [28a]. The first known tricarbonylchromium complex of an *ansa*-phane was **19a** and was prepared by Misumi [29]. Nearly all members have now been complexed with $Cr(CO)_3$ [30] and $Mo(CO)_3$[31] (Fig. 4).

The mono(tricarbonylchromium) complexes of [2_2]cyclophanes are all known. Cram [32] obtained the first members of the series in 1960: he complexed **8** as well as its higher homologs with longer bridges and obtained the parent complex **28** in 71% yield. Langer, Lehner and Schlögl prepared the

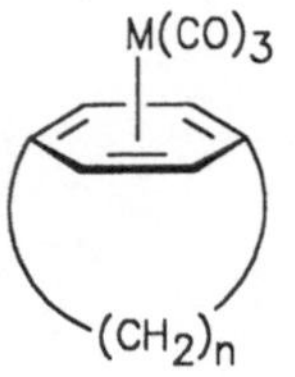

		uncomplexed	M = Cr	M = Mo
$M(CO)_3$	(n = 5)	17	–	–
	(n = 6)	18	18a	18b
	(n = 8)	19	19a	19b
	(n = 9)	20	20a	20b
$(CH_2)_n$	(n = 11)	21	21a	21b
	(n = 12)	22	22a	22b
	(n = 15)	23	23a	23b

Fig. 4. Mono(tricarbonyl) complexes of *ansa*-phanes with group VIb-elements [28b, 29–31]

complexes **25** [33], **26** [34] and **27** [35] of the isomeric *ortho-*, *meta-* and *metapara*-cyclophanes **24**, **6**, and **7** (Fig. 5). The *meta*-substituted phenylene ring in **7** is selectively complexed [35]. Mo(CO)₃ complexes of **6** and **8** (**29** [31] and **30** [31]) as well as the W(CO)₃ complex **31** [36] are known (Fig. 5).

Examinations on substituted cyclophanes indicate, that electron releasing substituents favor, and electron attracting substituents disfavor complexation [37].

All complexes are yellow crystalline solids which are soluble in most organic solvents. In solution they tend to decompose, complexes with tungsten being less stable than complexes with molybdenum and chromium [36]. The same trend is reflected in thermal stability and sensitivity towards oxygen, which is the reason why preparation of Mo(CO)₃- and W(CO)₃-complexes have to be carried out in lower boiling point solvents [23].

Cram succeeded in the first preparation of the bis(tricarbonylchromium) complexes of [2.2]paracyclophane (**8**) and higher homologs with longer distances between the benzene rings, by reacting the cyclophane with a second equivalent of Cr(CO)₆ [32]. The synthesis of **33** was first achieved by Misumi [29] and remarkably **24** seems to be inert towards biscomplexation even under drastic reaction conditions [33] (Fig 6a).

Higher members like [2₃](1,4)cyclophane (**34**) are available in a modified Wurtz coupling reaction [38]. Cram [32] described the complexation of only one benzene ring of **34**, while Elschenbroich [39] succeeded in the complexation of the second and third benzene ring by variation of stoichiometry and prolonged reaction time (Fig. 6b).

The two benzene rings in compounds like **8** lie on top of each other and thereby exhibit interesting electronic and spectroscopic interactions. Misumi [40] was successful in preparing multilayered cyclophanes with up to six benzene decks. Nowadays the mono- and bis(tricarbonylchromium) complexes **35a** and **35b** of the triple layered (**35**) as well as **36a** and **36b** of the quadruple layered *para*-cyclophane (**36**) are known (Fig 6c) [29].

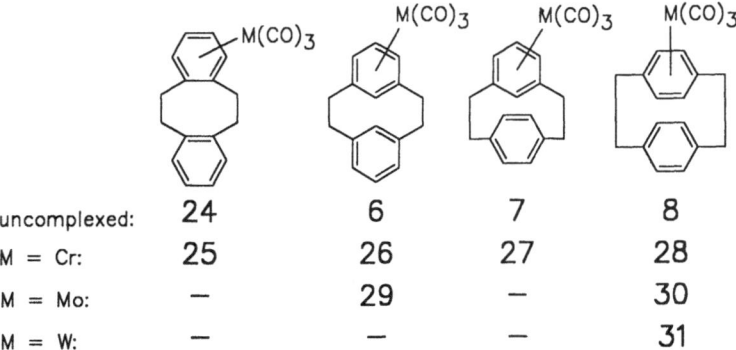

Fig. 5. Mono(tricarbonyl) complexes of [2.2]cyclophanes with group VIb-elements [31–36]

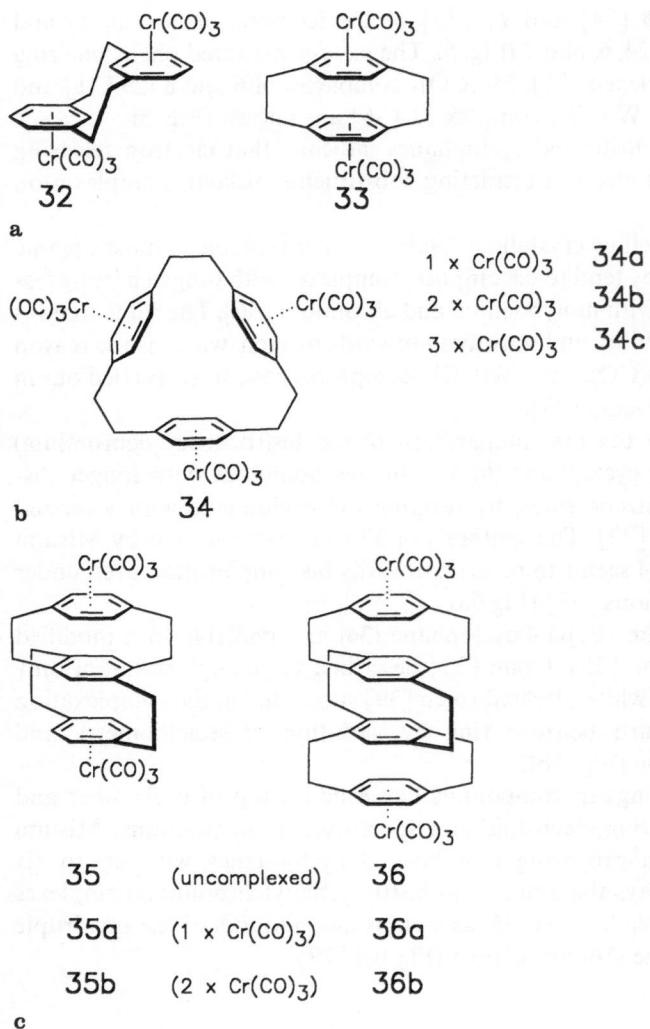

Fig. 6a–c. Bis- and tris(tricarbonylchronium) complexes of a. [2.2]cyclophanes [29, 34b], b. [2.2.2]-(1,4)cyclophane (34) [39] and c. triple and quadruple layered cyclophanes [29]

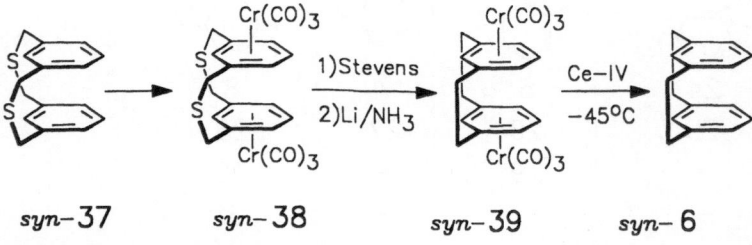

Fig. 7. Synthesis of syn-[2.2]metacyclophane [41]

All bis- and tris(tricarbonylchromium) complexes are yellow solids, which are not easily soluble in common organic solvents [29, 39].

$[2_2](1,3)$cyclophane (6) exists in two conformations, the stable *anti*-form, having stepwise-arranged benzene rings, and the *syn*-form, which above $0\,°C$ rearranges to the *anti*-form; the planes of the two benzene rings form an angle of

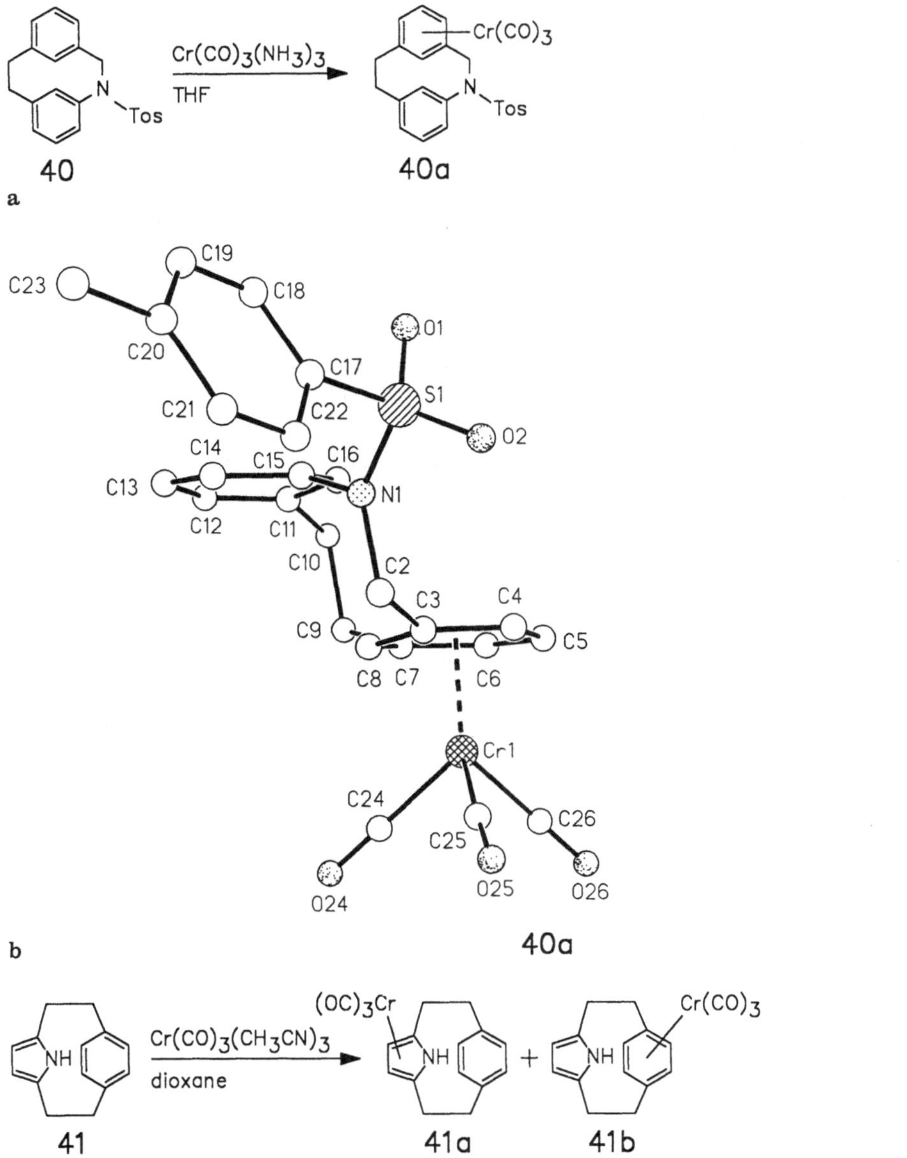

a

b
40a

c

Fig. 8a–c. Synthesis of the first Cr(CO)₃-complex of a. complexation of a heteracyclophane [45] and b. its X-ray structure [45] and c. complexation of a heteracyclophane [46]

29° [41a]. The synthesis of *syn*-6 by Mitchell [41] is an excellent example of using the chemical properties of the Cr(CO)$_3$-group for the preparation of compounds otherwise not yet available.

It is known that electron attracting substituents stabilize the *syn*-conformation of [3$_2$](2,11)dithiametacyclophane (37) [42], so that complexation with tricarbonylchromium yields almost exclusively *syn*-38 [41]. Stevens rearrangement followed by reduction with Li/THF at low temperatures leads to the formation of 39 and after decomplexation with Ce-IV/CH$_3$CN at −45 °C to *syn*-6 finally (Fig. 7) [41].

The possibility of extrusion of sulfur from thiaphanes, when exposed to transition metal carbonyls, is known [43]. 38 was the first transition metal complex of a dithiacyclophane [41].

Formal replacement of a carbon atom in the ethano bridge of 6 by a nitrogen atom gives the heteraphane 40 [44]. Complexation with Cr(CO)$_3$(NH$_3$)$_3$ in THF yields regioselectively 40a, where the chromium atom is bound to the one benzene ring not attached to the NTos-group [Tos = − SO$_2$(p − C$_6$H$_4$)CH$_3$]; (Fig. 8a) [45].

Atwood [46] obtained the first η^5-coordinated transition metal cyclophane complex (41a), reacting 41 with Cr(CO)$_3$(CH$_3$CN)$_3$ in dioxane (Fig 8c). The different complexation behaviour of 40 and 41 reveals, that in case of complexation with the Cr(CO)$_3$-group steric effects are more important than electronic reasons [45]. The Tos-group is placed above the benzene ring to which it is attached and shields it from complexation (Fig 8b) [45].

3.1.3 Complexes of Cyclophanes Containing *anti*-Aromatic π-Decks

A new group of [2$_2$]cyclophanes has only been available since recently, the benzene rings being replaced by *anti*-aromatic units such as the cyclobutadienyl moiety. Adams [47] used the stabilizing effect of the Fe(CO)$_3$-group towards cyclobutadiene [23a, 48] and achieved the synthesis of the complex 44, the smallest cyclophane with a metal stabilized *anti*-aromatic π-deck (Fig 9). Each of the two cyclobutadienyl rings of the chained bis(iron) complex 42 is formylated and cyclized to 43 via the McMurry reaction; reduction yields 44 in 23% yield with regard to the starting material 42 (Fig 9) [47].

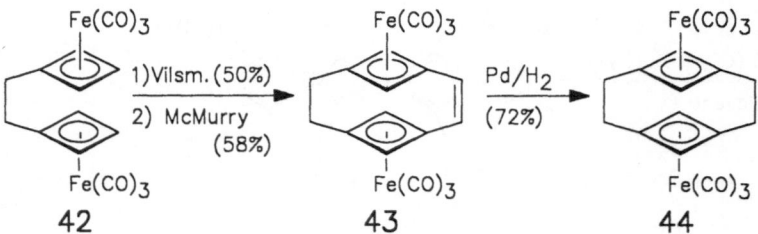

Fig. 9. Synthesis of the smallest cyclophane with metal stabilized *anti*-aromatic π-decks [47]

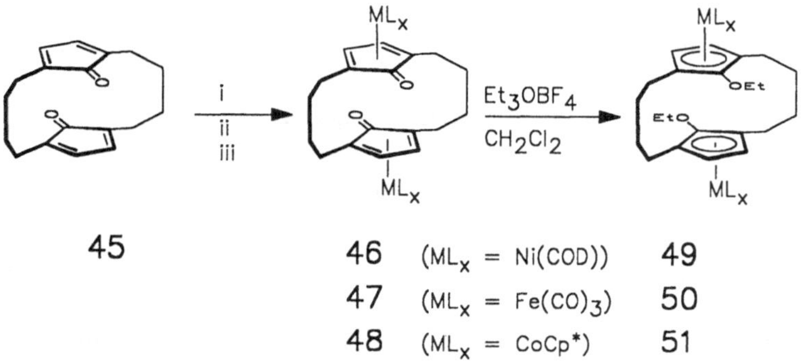

Fig. 10. Complexation of a cyclophane containing cyclopentadienone rings [49]; i) $Ni(CO)_4$ (85% yield); ii) $Fe_2(CO)_9$ (65% yield); iii) $Cp*Co(C_2H_4)_2$ (68% yield)

Jutzi [49] prepared the cyclophane **45**, which contains two doubly bridged cyclopentadienone rings. $Ni(COD)_2$, $Fe_2(CO)_9$ or $Cp*Co(C_2H_4)_2$ as complexing agents lead to complexation of both π-decks in good yields (Fig 10). The cyclopentadienone rings in the complexes **46–48** exhibit the anti-conformation and are planar. Addition of strong acids, like CF_3SO_3H or HBF_4, as well as Lewis acids like $Et_3O^+BF_4^-$, lead to the formation of cyclopentadienyl rings under preservation of the coordinated metal fragments (Fig. 10) [49].

3.2 Sandwich Complexes

3.2.1 General Remarks

The interactions between the aromatic π-decks in [2$_n$]cyclophanes was investigated extensively by photoelectron and electron spin resonance spectroscopy [50]. It was revealed that the decks can be regarded as a single electron system [50]. This feature is also found in multilayered cyclophanes [6, 50].

Interesting electrical properties are to be expected with the stepwise extension of this π-system. The preparation of multilayered cyclophanes proved to be laborious [6]; nevertheless new synthetic methods in transition metal chemistry of arenes have opened up a promising alternative approach via preparation of multidecker sandwich complexes (structure type D in Fig. 3). First row transition metals like chromium, iron and cobalt [51] form strong coordinative bonds with arenes when their oxidation state is low [48a] whereas second and third row elements like ruthenium, rhodium and iridium are strongly bonded towards arenes in higher oxidation states [48a, 51]. Sandwich complexes of cyclophanes can be divided into two groups;

a) the metal atom is included in the "cavity" of the cyclophane (interannularly bridged ligands)

b) the metal atom is coordinated at the outer face of the cyclophane (intra-annularly bridged ligands)

3.2.2 Sandwich Complexes with Interannularly Bridged Ligands

3.2.2.1 Inclusion Complexes Containing Chromium

In 1969, Timms [52] prepared di(benzene)chromium (**52**) through metal ligand cocondensation. The method is of general application and leads to the transfer of chromium atoms to arenes, resulting in sandwich complexes.

In 1978 Elschenbroich [53] applied this method to the inclusion of a chromium atom into the cavity of **8**; **53** was formed (besides **54** and polymeric products) in 5% yield (Fig 11). The interarene distances in **8** [54] are shorter compared with **52** [55], but are obviously long enough for the inclusion of a chromium atom. **53** therefore probably constitutes a compressed sandwich complex [53]. Extending each bridge in **8** by one methylene group leads to [3$_2$] (1,4)cyclophane (**55**), where the interarene distance is substantially longer [56] than in **8** but, with the strong π-π-interaction still being present [56a]. Thus preparation of **10** was achieved via cocondensation in 2% overall yield (Fig. 12) [15]. The complex was purified by oxidation to the radical cations **56** and **57** [15].

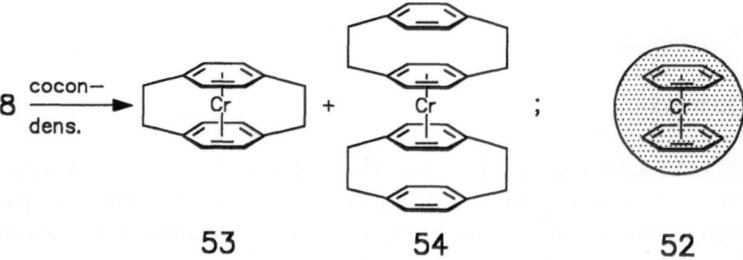

Fig. 11. Cocondensation products of chromium metal vapor and [2.2]paracyclophane (**8**) [53]

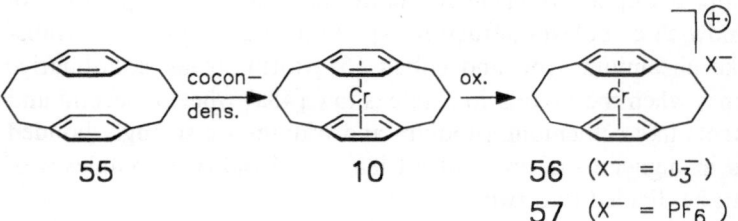

Fig. 12. Cocondensation product of chromium metal vapor and [3.3]paracyclophane (**55**) and its oxidation to the corresponding radical cation **56** and **57** [15]

In 1986, all methylene groups in **8** were replaced by heteroatoms for the first time leading to its Si-analog **58** [57]. Due to the low availability of **58**, a direct preparation via cocondensation was not promising; a template reaction was used instead and gave **59** in 3.5% yield after reductive dechlorination of the sandwich complex **61** (Fig. 13) [58].

The uncharged complexes **10, 53** and **59** are thermally stable solids. **53** and **59** were separated from the byproducts by sublimation and oxidation to their radical cations [15, 53]. The radical cation of **53** is more stable towards solvolysis in comparison to other open-chained di(benzene)chromium derivatives and is even resistent towards HCl for one week [53]. In contrast, the radical cation of **59** is sensitive towards solvolysis [58].

Replacement of carbon atoms in the propylene bridge of the complex **56** by heteroatoms, particularly by metal atoms as is represented by nickel in complex **65** [20], is interesting in view of the redox chemistry. Preparation of **65** succeeds also via a template reaction (Fig 13).

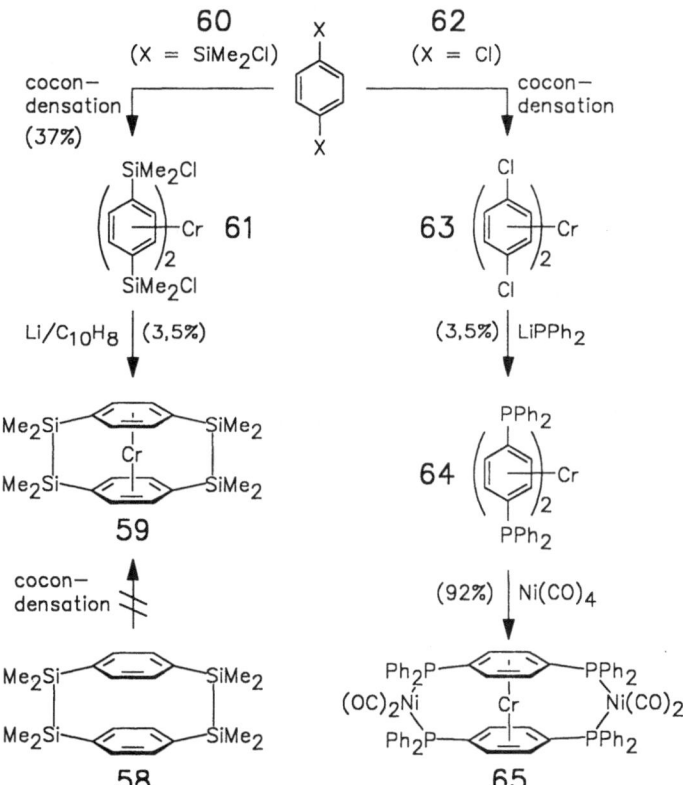

Fig. 13. Synthesis of cyclophane complexes containing heteroatoms in the bridges with η^{12}-coordinated chromium metal [20, 58]. Reaction **58** → **59** has not been carried out.

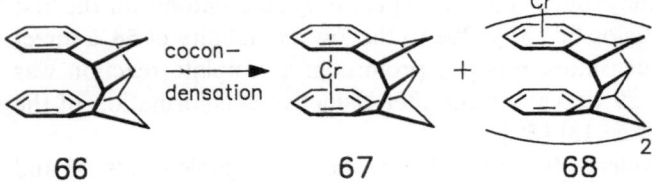

Fig. 14. Cocondensation products of chromium metal vapor and a [3.3]orthocyclophane [59]

The phane **66** can be regarded as a threefold additionally bridged [3$_2$] (1,2)cyclophane; via cocondensation a chromium atom can be captured between the two benzene rings lying "face to face" with each other [59]. The complex **68**, where two cyclophane molecules are coordinated to a chromium atom, is a byproduct (Fig 14) [59].

3.2.2.2 Inclusion Complexes with Silver

Pierre [60] observed a considerable increase in solubility of Ag-trifluoromethanesulfonate (triflate) in chloroform, when **34** was added. An analogous open chained reference compound does not show this effect. Cyclophanes which include metal cations due to π-interaction with arene moieties are called "π-prismands" [60]. However the silver ion is not situated symmetrically between the three benzene rings, but on the pseudo threefold axis a little outside the plane defined by the three ethano bridges [61]. "*Deltaphane*" (**69**) contains three further ethano bridges, which leads to even greater rigidity, and is nearly insoluble in THF, but dissolves immediately on addition of silver triflate [62]. Dynamic NMR experiments revealed that the silver ion is exchanged intermolecularly and that it does not channel the cavity of **69** (Fig. 15) [62].

The silver-carbon-bond vectors in **69a,b** are oriented nearly perpendicular to the benzene ring planes, which primarily infers participation of their π-electron systems at the silver–carbon bond [62]. This kind of bonding is somewhat different to those found in silver ion complexes of other arenes [63].

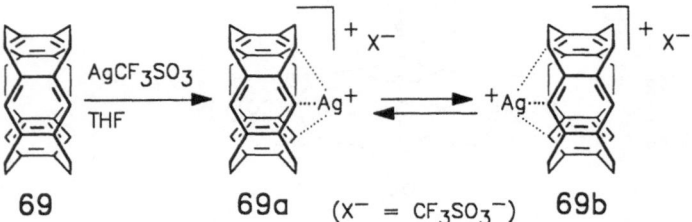

Fig. 15. Intermolecular exchange of silver cation in the complex with *deltaphane* [62]

3.2.2.3 Inclusion Complexes with Iron: Ferrocenophanes (fcp)

The building block of these phanes, ferrocene (fc, **70**), was first prepared in 1951 [64] and represents the first iron-containing hydrocarbon. It shows distinctive "aromatic" character and reacts easily with Friedel-Crafts reagents [8, 65].

The syntheses of fcp differ from those of other transition metal complexes of cyclophanes and starts with **70** itself [8, 65]. Only ethano bridged fcp [66] have to be prepared from the free ligand and iron-II-ions. A typical reaction sequence is illustrated in Fig. 16 [67]: **72** is reduced to **73** using LAH/AlCl$_3$. Ring extension [67, 68] with diazomethane followed by reduction yields **75** (Fig. 16) [67].

Direct introduction of a C$_4$-bridge is unsuccessful, because upon acylation of γ-butyric acid homoannularly cyclization selectively occurs [68]. This is why C$_4$- and C$_5$-bridges have to be constructed by stepwise ring extension reactions [68].

After Lüttringhaus [69] had prepared the first bridged fcp (**4**), a main emphasis in fcp-chemistry was to capture the iron atom completely between the Cp-rings, through further bridging. Hisatome [70] succeeded in preparing the first fully bridged fcp **76**. Later on [71], the same group achieved symmetrical bridging by five C$_4$-bridges; in following Boekelheides "superphane" [4b] the

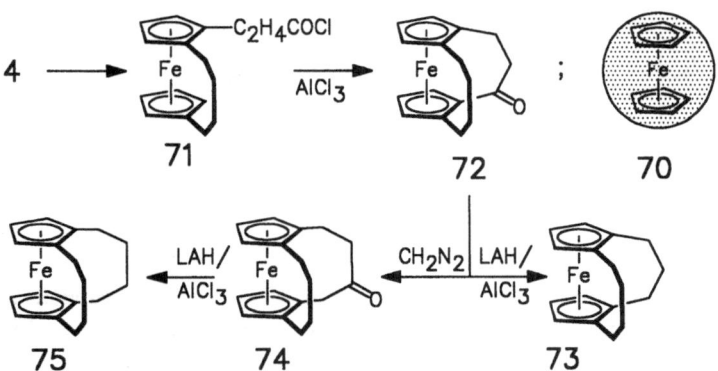

Fig. 16. Synthesis of trimethylene and tetramethylene bridges in ferrocenophanes (fcp) [67]

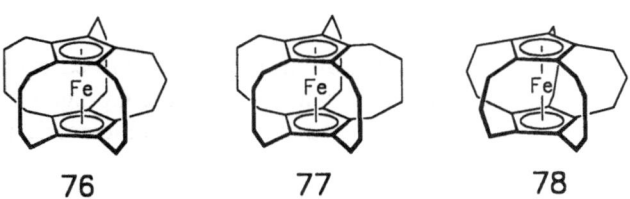

Fig. 17. Some fully bridged ferrocenophanes [70, 71]

COCl
Fe
O
Fc
Fc
O
Fc
O
Fc ≡
Fe
79
80

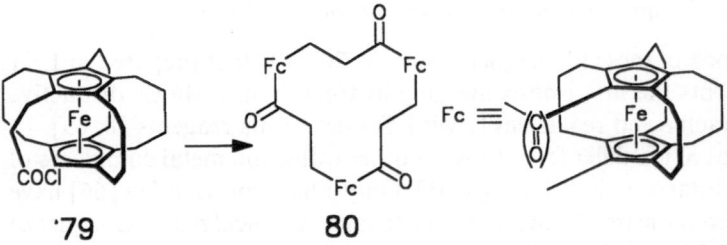

Fig. 18. Synthesis of a fcp (**80**) containing inter- and intramolecular bridges [74]

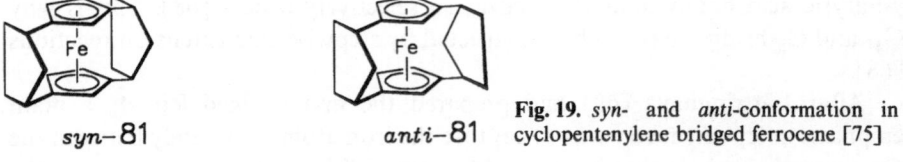

Fe
Fe
syn–81
anti–81

Fig. 19. *syn*- and *anti*-conformation in cyclopentenylene bridged ferrocene [75]

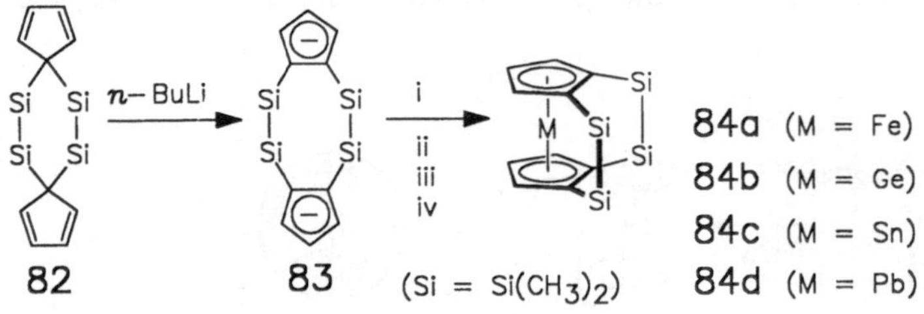

Si
Si
Si
Si
n–BuLi
i
ii
iii
iv
M
84a (M = Fe)
84b (M = Ge)
84c (M = Sn)
84d (M = Pb)
82
83
(Si = Si(CH3)2)

Fig. 20. Inclusion complexes of silicon bridged metallocenes (i–iv) MCl_2 (M = Fe, Ge, Sn, Pb) [77]

complex was called "[4]superferrocenophane" (Fig 17) [71]. The respective "[3]- and [5]superferrocenophanes" remained undiscovered so far, however, **78** containing four C_3-bridges and one C_4-bridge [72], as well as an fcp containing four C_5-bridges have recently been described (Fig. 17) [73].

The interesting molecule **80** was formed as the main product in an attempt to obtain **77** via cylization of the β-propionic acid derivative **79** (Fig. 18) [74]. **80** exhibits inter- as well as intraannularly bridged phane structure und includes two molecules of chloroform as guests inside the cavity [74].

An alternate approach for capturing an iron atom in a hydrocarbon framework was developed by Astruc and Batail, who bridged the Cp-ligands with (1,3)cyclopentandiyl units [75]; **81** exists in the *syn*- and *anti*-conformation (Fig. 19).

Only recently have Si-bridged Cp-ligands become known [76]. The synthesis of **84a** was achieved by reacting **82** with two equivalents of *n*-BuLi and then $FeCl_2$ (Fig. 20) [77]. The silatropic rearrangement of **82**, accompanied with deprotonation, is a typical feature of the chemistry of Si-substituted cyclopentadienes [76, 77]. In contrast to C_2-bridges, Si_2-bridges are long enough for depositing an iron-II-ion (and analogously germanium-II, stannium-II- und lead-II-ions) between the two Cp-rings; compounds **84b–d** represent the first examples of doubly bridged metallocenes with p-block elements (Fig. 20) [77].

3.2.3 Sandwich Complexes Containing Intraannularly Bridged Ligands

The number of complexes of this type is considerably higher than for inclusion- and carbonyl complexes, because the components of the complexes can be varied much more: Substituted cyclophanes from [2₂]- up to [2₆]cyclophanes, as well as indenophanes are used as ligands for metal fragments, which may consist of chromium, iron, ruthenium, rhodium, iridium or cobalt as metal units, and Cp-, Cp*- and C_6R_6-units as stabilizing co-ligands. From Fig. 21 the importance of the oxidation state and position in the periodic table of the individual metals can be seen [48a]: cobalt-III forms monocomplexes only, whereas the lower sited iridium-III also forms the biscomplexes; for cobalt,

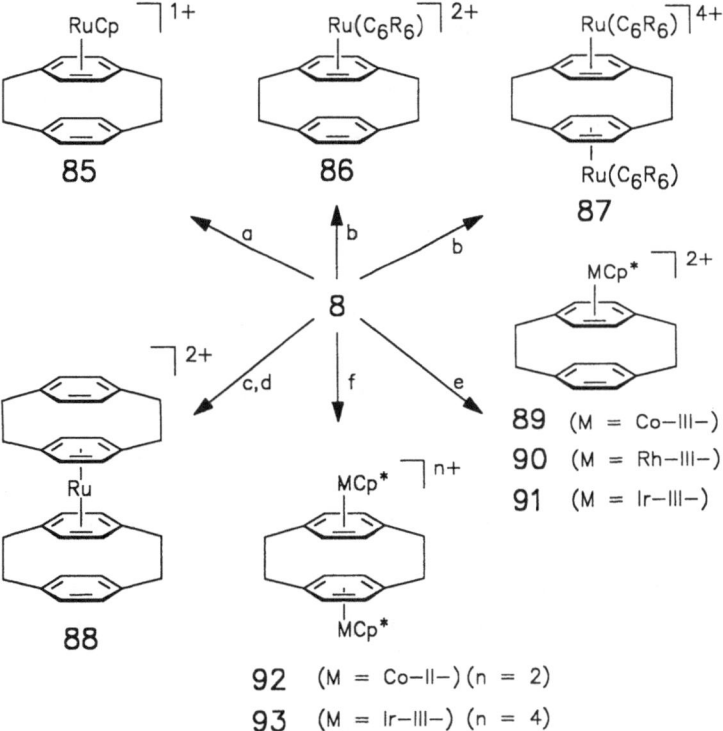

Fig. 21. Reaction paths (a to f) to multidecker sandwich complexes of ruthenium, cobalt, rhodium and iridium with [2.2]paracyclophane (**8**) as ligand

reduction to cobalt-II is necessary to achieve biscomplexation [78]. Stabiliz-ation with the cycopentadienyl-(Cp) or pentamethylcyclopentadienyl-(Cp*) ligand is essential [78]. The same trend is seen in the iron triade: whereas $(\eta^6\text{-}$arene$)_2$Fe^{2+} is unstable and the $(\eta^6\text{-}$arene)Fe^{2+} fragment has to be stabilized with a Cp- or Cp* -ligand [79], ruthenium-II in contrast forms air stable $(\eta^6\text{-}$arene$)_2$Ru^{2+} complexes, due to stronger π-backbonding abilities [80]. Com-plexes of structure type D (Fig. 3) are instable for most metals and known mainly for ruthenium.

3.2.3.1 Ruthenium, Cobalt, Rhodium, and Iridium Complexes

Examplified with **8** as ligand, the possible multidecker sandwich complexes are depicted in Fig. 21. Only four sandwich complexes of ruthenium-II are yet known for *ansa*-phanes [81].

The preparation of these complexes has made necessary the development of new complexing agents, which transfer the respective metal moiety to the cyclophane. Generally three coordinating sites of the metals are bound to labile solvent molecules. These are easily replaced thermally by the cyclophane (Fig. 22).

Monomeric building units of structure type D in this series (Fig. 3) are known only for ruthenium (complex **88** in Fig. 21). Method c [82b, 90] is particularly well suited for the preparation of such complexes, whereas method d [83] is limited to the few cyclophanes that can be reduced via Birch's method (Fig. 22). The halogen dimers $[(\eta^5\text{-}Cp^*)MX_2]_2$ are known (X = Cl, I; M = Co-III-[84], Rh-III-[85], Ir-III-[86]) and are easily transferred into the active carriers **101–103** (method e) [78]. The Co-II-carrier (method f) [78] is a stronger complexing agent than the respective Co-III-carrier and effects biscomplexation also (Fig. 22).

3.2.3.2 Iron Complexes

The classic method of preparing bis$(\eta^6\text{-}$arene)iron^{2+} complexes by reacting the arene with FeCl$_2$/AlCl$_3$ is not directly transferable to cyclophanes, because they, even in the presence of traces of HAlCl$_4$, isomerize [91]. However, after addition of Me$_3$Al$_2$Cl$_3$ (method g) as a proton scavenger the bisiron complex **107** is formed [79] as an orange solid which is sensitive towards oxidation in solution and in the solid state (Fig. 23). Methyl substituted cyclophane ligands provide stabilization to some extent [92].

Photolyses of $(\eta^5\text{-}Cp)$Fe$(\eta^6\text{-}$arene) cations (method h) [94–96] lead to de-composition into iron^{2+} and ferrocene when suitable ligands are absent [93]. Cyclophanes have proved to be stronger π-donors than open-chained benzene analogs, e.g. *p*-xylene, and therefore $(\eta^5\text{-}Cp)$Fe$(\eta^6\text{-}(p\text{-}$xylene$))$ is a strong com-plexing agent for the CpFe^{1+}-moiety (method h) [94–96]. The same transfering unit can be obtained by reaction of AlCl$_3$/Al-powder with ferrocene (method i) [97] (Fig. 23).

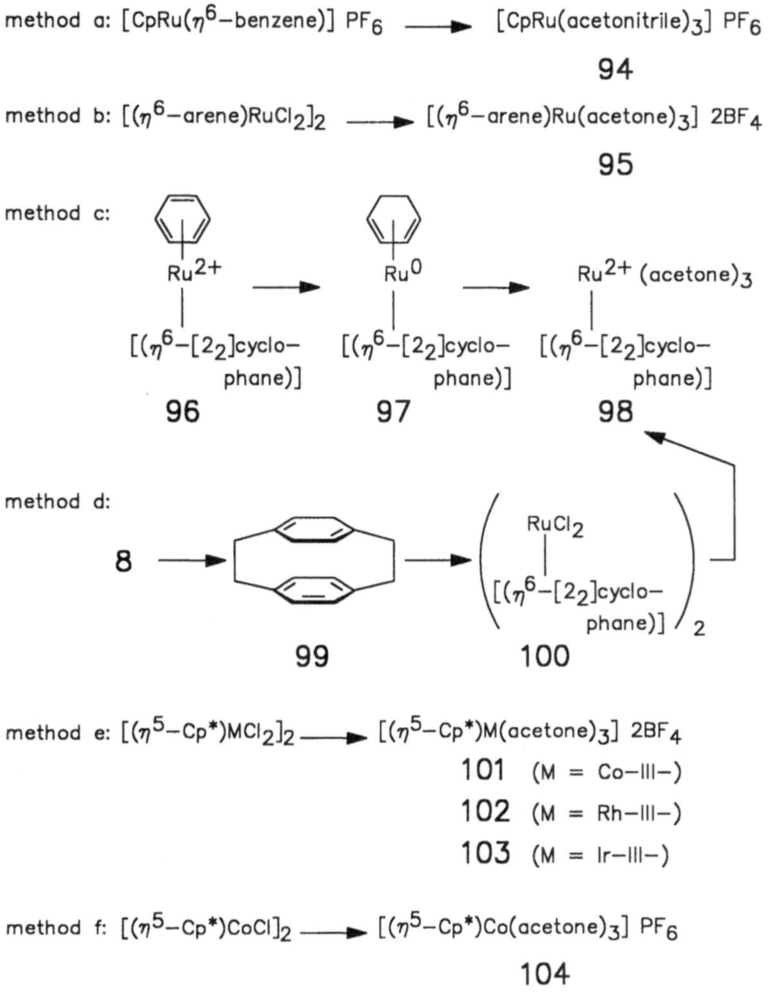

method a: $[CpRu(\eta^6-benzene)] PF_6 \longrightarrow [CpRu(acetonitrile)_3] PF_6$

94

method b: $[(\eta^6-arene)RuCl_2]_2 \longrightarrow [(\eta^6-arene)Ru(acetone)_3] 2BF_4$

95

method c:

$Ru^{2+} \longrightarrow Ru^0 \longrightarrow Ru^{2+} (acetone)_3$

$[(\eta^6-[2_2]cyclo-$ $[(\eta^6-[2_2]cyclo-$ $[(\eta^6-[2_2]cyclo-$
phane)] phane)] phane)]

96 **97** **98**

method d:

$8 \longrightarrow 99 \longrightarrow 100$

$\left(\begin{array}{c} RuCl_2 \\ | \\ [(\eta^6-[2_2]cyclo- \\ phane)] \end{array} \right)_2$

99 **100**

method e: $[(\eta^5-Cp^*)MCl_2]_2 \longrightarrow [(\eta^5-Cp^*)M(acetone)_3] 2BF_4$

101 (M = Co–III–)

102 (M = Rh–III–)

103 (M = Ir–III–)

method f: $[(\eta^5-Cp^*)CoCl]_2 \longrightarrow [(\eta^5-Cp^*)Co(acetone)_3] PF_6$

104

Fig. 22. Synthesis of some reagents provided with a reactive metal fragment (reaction paths a [87], b [88, 89], c [82b, 90], d [83], e [78] and f [78])

Among *ansa*-phanes only one iron complex is known [98]. Far more common and more stable are CpFe-complexes of phanes in which the iron atom is η^5-coordinated to the cyclophane. This is realized in fcp, in which the Cp-ligand is either stapled [18, 33, 99], or in indenophanes [103] in which the Cp-ring is annellated to a benzene ring. Two main approaches for the preparation of fcp containing intraannularly bridged Cp-ligands, are known:

- derivatization of ferrocenes already possessing phane structure
- deprotonation of cyclopentadiene-units and subsequent reaction with FeCl$_2$

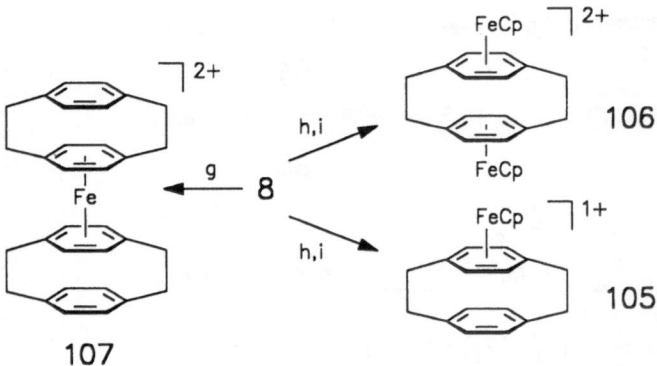

method g: FeCl₂/AlCl₃/Me₃Al₂Cl₃

method h: [(η⁵–Cp)Fe(p–xylene)]/CH₂Cl₂/hν

method i: (η⁵–Cp)₂Fe/AlCl₃/Al

Fig. 23. Reaction paths (*g* to *i*) to cyclophane complexes with iron [79, 94–97]

The formylated ferrocene **108** reacts with **109** to give the aldol-condensation product **110** (Fig. 24) [18]; the bridged Cp-rings possess phane structures itself, but the nomenclature refers to the bridged ferrocene nucleus [13, 33]. The ketone function can be reduced first, followed by hydrogenation of the C-C-double bond or vice versa to give **111** (Fig. 24) [18].

Shortening the bridges connecting the two cyclopentadiene rings with one another is realized in compound **112** (Fig. 25) [99]. After simple deprotonation preparation of heterobimetallic complexes such as **116** is possible [100], or alternatively double deprotonation of **115** provides access to one-dimensional polymeric metal complexes (Fig. 25) [100, 101].

The dianion of **112** yields the chromocene **117** in 43% yield as a dark red and air sensitive solid, by reacting it with CrCl₂ (Fig. 25) [102a]. This alternate

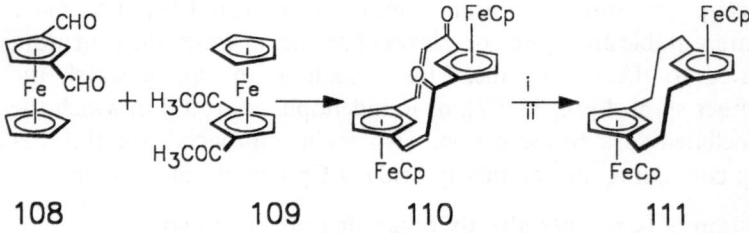

Fig. 24. Aldole condensation of two ferrocene units; the fcp **110** can be reduced with (i) PtO₂/H₂ first followed by reduction with (ii) LiAlH₄/AlCl₃ or vice versa [18]

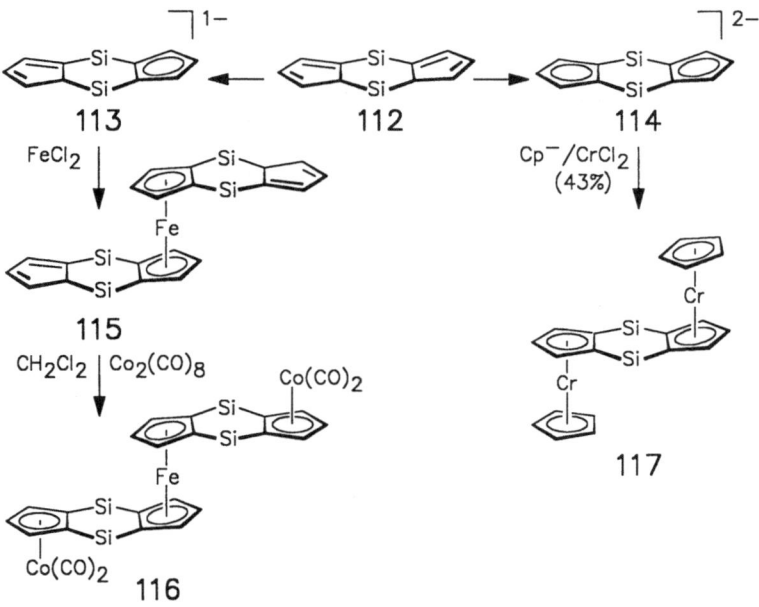

Fig. 25. Some metallocenophanes containing silicon bridges and the heterobimetallic complex **116**: an alternate approach to an organometallic polymer [99–101]

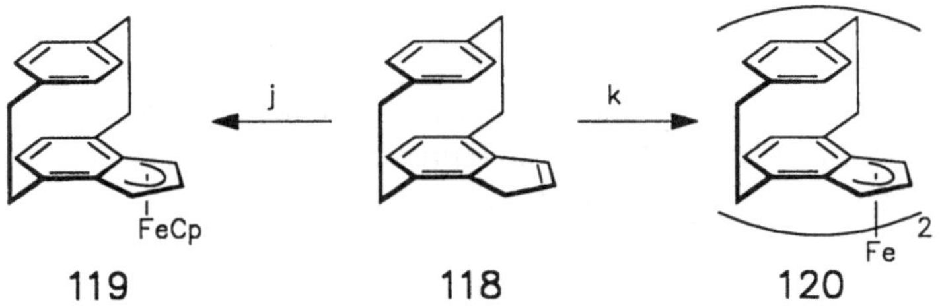

method j: MeLi/THF/CpLi/FeCl₂

method k: Na/THF/FeCl₂

Fig. 26. Iron complexes of an indenophane [103, 105]

approach offers the construction of multi-layered metallocenes with metals other than iron [102b,c, 103].

The method for preparation of ferrocene (**70**) [104], applied to indeno-phanes, links the two most extensively investigated two-layered molecules in organic and metal organic chemistry, [2₂](1,4)cyclophane (**8**) and ferrocene (**70**) respectively (Fig. 126) [103, 105].

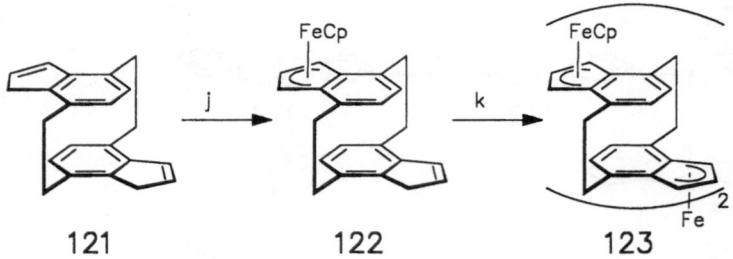

Fig. 27. The first oligomeric unit (**123**) of an organometallic polymer containing ferrocene- and [2.2]paracyclophane units (reaction paths *j* and *k*) [103, 105]

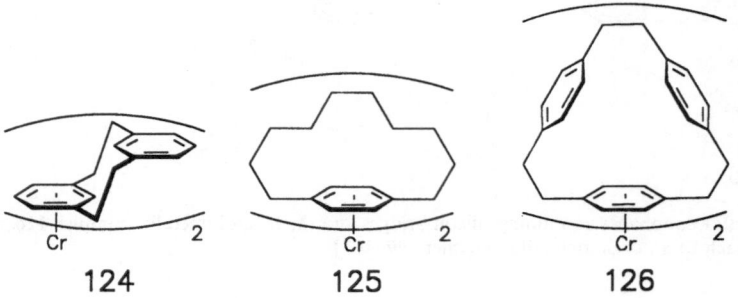

Fig. 28. Cyclophane complexes of $2 \times \eta^6$-coordinated chromium atom [39, 107, 108]

Hopf [106] succeeded with the synthesis of the indenophane **121**, which gives **123** in a two step procedure (Fig. 27) [106]. **123** represents a first oligomeric part of a polymer of structure D (Figs. 3 and 27) [103, 105].

3.2.3.3 Chromium Complexes

The sandwich complexes **125** [107], **124** [108] and **126** [39] were prepared via cocondensation from the free ligands **1**, **6** and **34** respectively; they are orange-brown or yellow-brown sublimable solids (Fig. 28).

3.2.3.4 Silver and Copper Complexes

Arene complexes with mint metals are rather rare. The arene metal bond is weaker compared to the other transition metals and hapticity is lower. For this reason the stronger π-donor properties of the cyclophanes compared to open chained arenes [109c] make the preparation in benzene solution possible. The cyclophane is simply heated with silver-I (Fig. 29a) [109a] or copper-I (Fig. 29b) [109b] and yields the respective complex formed in almost quantitative yield. It is noteworthy that the silver ion in **128** shows no interaction with the benzene ring at all; the naphthalene unit acts as π-donor only [109a].

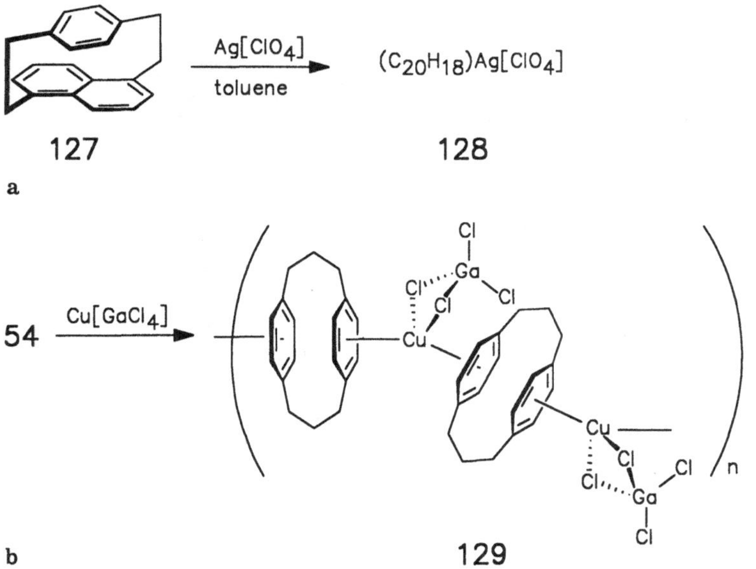

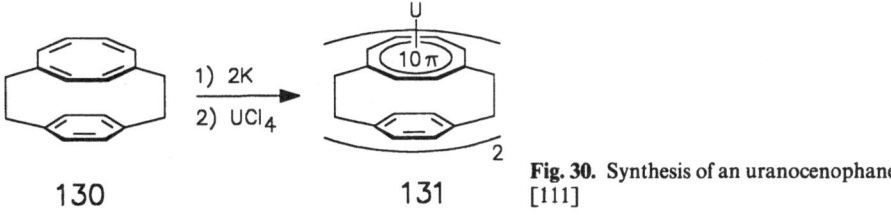

Ag[ClO4]
toluene
(C20H18)Ag[ClO4]
127
128
a
54
Cu[GaCl4]
Cl
Ga
Cl
Cl
Cu
Cu
n
b
129

Fig. 29a. A silver complex of a naphthaleno cyclophane [109a] and **b.** a copper complex of [3.3]paracyclophane [109b]

U
10π
1) 2K
2) UCl4
2
130
131

Fig. 30. Synthesis of an uranocenophane [111]

3.2.3.5 Uranium Complexes

Cyclophanes having a cyclooctatetraene deck like **130** can be reduced to the dianion, which, analogous to the preparation of uranocenes [110], gives the uranocenophane **131** after reaction with UCl_4 (Fig. 30) [111]. Because of the low stability a full characterization has not been possible, but the green colour and the proton NMR-spectrum is in agreement with the structure of **131** given [111].

3.2.4 Sandwich Complexes of Cyclophanes Containing *anti*-Aromatic π-Decks

Gleiter [112] recently opened up access to a new group of "superphanes" containing *anti*-aromatic π-decks, which are stable only when coordinated to a transition metal. Reaction of cyclodiynes (**132** and **134**) with $CpCo(CO)_2$ gives the intermediates **135** and **137** respectively, which yield the CpCo-stabilized "superphanes" **138** [113] and **140** [114] respectively after addition of a second equivalent of $CpCo(CO)_2$ (Fig. 31a). Remarkably, under analogous conditions,

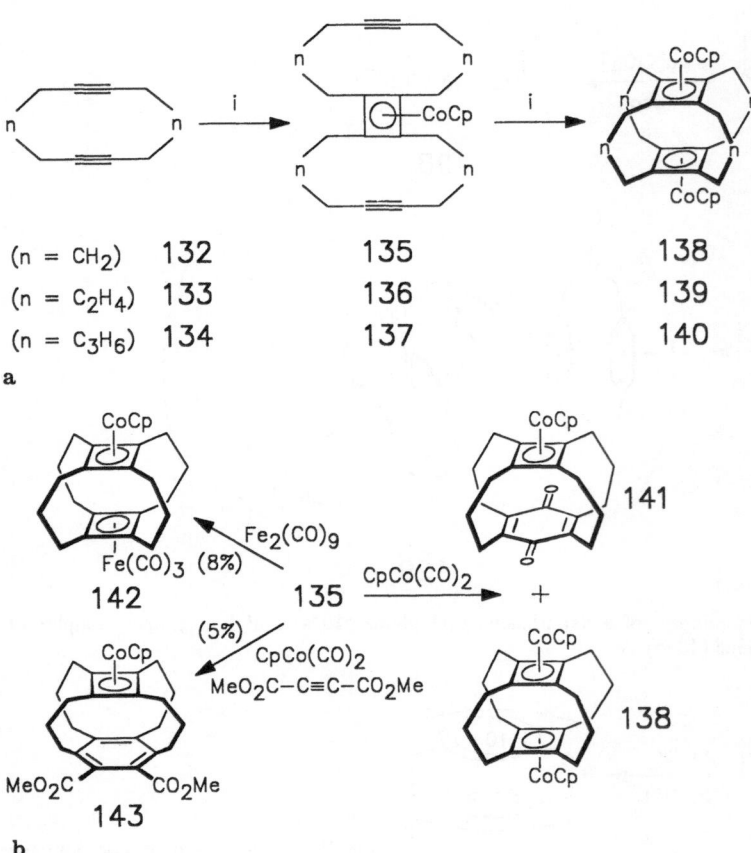

a

b

Fig. 31a. Reaction path to a new class of superphanes containing cobalt stabilized *anti*-aromatic π-decks and **b**. **135** as a useful intermediate for the preparation of various cyclophane complexes [112–114]

133 does not lead to the formation of the "superphane" **139** [112], which remains yet unknown.

A generalization of this reaction path is possible, when the intermediate **132** is trapped and reacted with various other transition metal moieties, some examples of which are illustrated in Fig. 31b [114].

4. Properties

4.1 Geometrical Considerations

One characteristic feature of [2ₙ]cyclophanes is the distortion of their benzene rings into unsymmetrical boat forms; the degree of distortion is smaller but still recognizable in $[3_2](1,4)$cyclophane (**55**) [56]. For this reason it is interesting to examine the influence of a complexed metal on the degree of the distortion of the benzene rings.

Due to the short ethano bridge in **6** [115], the inner carbon atoms C-8 and C-16 come into close proximity and force the benzene rings to distort into shallow boat forms. Comparison of numerous tricarbonylchromium complexes [45, 116, 117], as well as an iron-[118a] and a ruthenium complex [90, 119] with the respective metal free ligands [115, 118a, 120] reveal, that the coordinated metal causes no distinct flattening of the benzene rings (Fig. 32). The small geometric changes observed should be examined with care because no clear tendency can be recognized and they rather seem to stem from packing effects in the crystal [90, 45, 117]. Flattening the benzene rings would bring the inner carbon atoms (C-8, C-16) into even closer proximity, which seems not possible to a noticable extent [117].

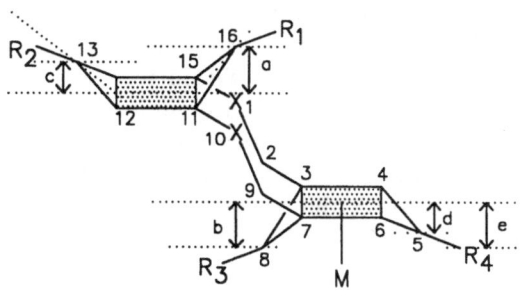

uncomplexed			complexed	
		$R_1 - R_4 = H/X = H$		
6			**26** (M = Cr(CO)$_3$)	
a = + 11.2 pm	c = + 5.0 pm		a = + 11.3 pm	c = + 2.6 pm
b = − 11.2 pm	d = − 5.0 pm		b = − 12.0 pm	d = − 4.3 pm
		$R_1, R_3 = H/R_2, R_4 = Me/X = H$		
144			**145** (M = FeCp)	
a = + 12.1 pm	c = + 4.4 pm		a = + 12.2 pm	c = + 5.2 pm
b = − 12.1 pm	d = − 4.4 pm		b = − 12.8 pm	d = + 1.6 pm
		$R_1, R_3 = Me/R_2, R_4 = H/X = H$		
146			**147** (M = RuMe$_6$)	
a = + 18.7 pm	c = + 8.8 pm		a = + 18.0 pm	b = + 9.0 pm
b = − 18.7 pm	d = − 8.8 pm		b = − 20.0 pm	c = − 2.0 pm
		$R_1 - R_3 = H/R_4 = SiMe_3/X = C(S_2(C_3H_6))$		
148			**149** (M = Cr(CO)$_3$)	
a = + 10.4 pm	c = + 5.1 pm		a = + 9.9 pm	b = + 5.3 pm
b = − 12.5 pm	d = − 5.7 pm		b = − 12.3 pm	d = − 2.1 pm
e = − 15.4 pm			e = + 20.0 pm	

Fig. 32. Influence of the transition metal on the geometry of [2.2]metacyclophanes

The distances between the metal atoms and the plane of the coordinated benzene ring are slightly increased compared to open chain analogs (chromium [116, 117], iron [118. 121]).

In the series of $Cr(CO)_3$-complexes of substituted arenes, there is considerable repulsive interaction between the $Cr(CO)_3$-group and an electron releasing group [122]. In the complex 149, the voluminous $SiMe_3$-group is situated (above the defined plane), but situated below it in the metal free ligand 148, impressively demonstrating repulsive steric interactions (Fig. 32) [117].

The two benzene rings in 8 lie directly on top of each other in close proximity, so that their π-clouds interpenetrate [54]. The $Cr(CO)_3$-group withdraws electron density from the complexed benzene ring [122] and therefore decreases the repulsive interaction between the two π-decks. De Meijere [123] in extensive X-ray structure investigations of chromium complexes of 8 [124] and some of its derivatives ascertained that the arene rings are considerably flattened (Fig. 33).

A decrease in interarene distances a and b is also observed, when a chromium atom is bound inside the cavity of the cyclophane (Fig. 34) [56, 58]. Replacement of a bridging carbon in 8 with silicon increases the interarene distances substantially; therefore inclusion of a chromium atom gives almost the same geometric parameters as were observed for 52 (Fig. 34) [56, 58]. No X-ray structure is known for 10, but from its radical cation 56, which shows decreased interarene distances compared to 55; the benzene rings are still considerably distorted into boat forms (Fig. 34) [54, 125].

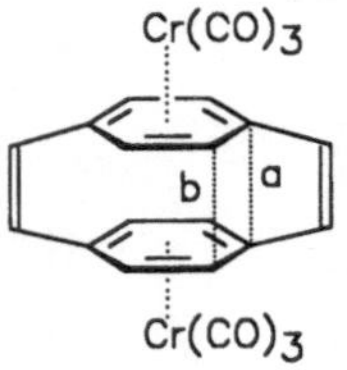

Cr(CO)₃

150 (uncomplexed) a = 280 pm b = 314 pm

151 (1 x Cr(CO)₃) a = 272 pm b = 305 pm

152 (2 x Cr(CO)₃) a = 268 pm b = 302 pm

Cr(CO)₃

Fig. 33. Influence of the $Cr(CO)_3$-group on the geometry of [2.2]paracyclophan-1,9-diene [123]

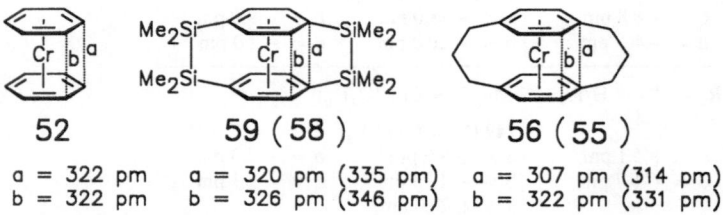

52

a = 322 pm
b = 322 pm

59 (58)

a = 320 pm (335 pm)
b = 326 pm (346 pm)

56 (55)

a = 307 pm (314 pm)
b = 322 pm (331 pm)

Fig. 34. Influence of a η^{12}-coordinated chromium atom on the geometry of [2.2]paracyclophanes (the structural data of the uncomplexed phane are given in brackets)

The complex **35a** shows two different distances (d_1 and d_2) between the planes of benzene rings, with d_1 being smaller than expected (Fig. 35) [126]. The phase transition of the crystal from monoclinic (− 95 °C) to triclinic (− 160 °C) is probably due to changes in molecular packing; the influence of the $Cr(CO)_3$-group shows the same tendency in both phases [126].

In multibridged phases rigidity increases and complexation e.g. with $FeCp^{1+}$ does not affect the conformation at all [118a, 127].

In cyclophanes containing *anti*-aromatic π-decks, interarene distances are considerably smaller than those containing aromatic π-decks (with the same bridge length). Therefore d is smaller in **44** [47] than in **33** [123], and smaller in **138** [113] than in **56** [125] (Fig. 36).

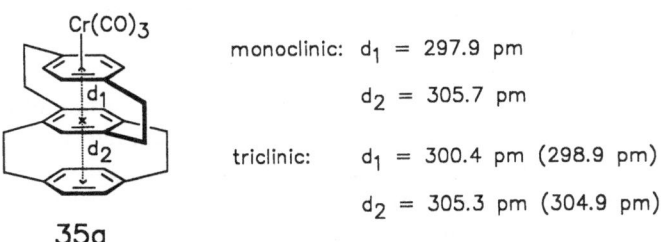

monoclinic: d_1 = 297.9 pm

d_2 = 305.7 pm

triclinic: d_1 = 300.4 pm (298.9 pm)

d_2 = 305.3 pm (304.9 pm)

35a

Fig. 35. Influence of the $Cr(CO)_3$-group on the geometry of multi-layered cyclophanes [126]

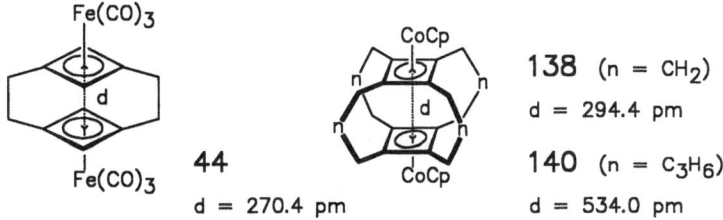

44

d = 270.4 pm

138 (n = CH_2)

d = 294.4 pm

140 (n = C_3H_6)

d = 534.0 pm

Fig. 36. "Interarene" distance in cyclophanes with metal stabilized *anti*-aromatic π-decks

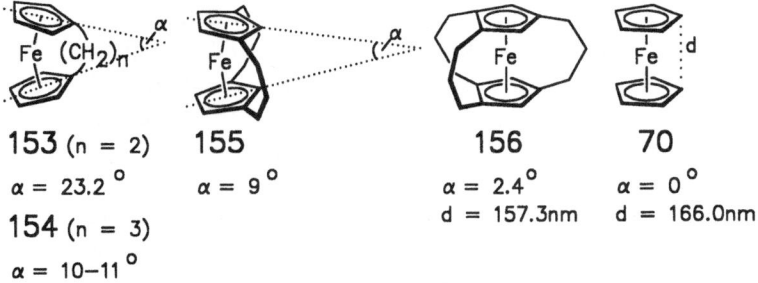

153 (n = 2)

α = 23.2 °

154 (n = 3)

α = 10–11 °

155

α = 9 °

156

α = 2.4 °

d = 157.3nm

70

α = 0 °

d = 166.0nm

Fig. 37. Influence of number and position of trimethylene bridges on the geometry of fcp [129b,c]

Interannular bridging of ferrocene (70) has strong consequences, because the short trimethylene group leads to compression, and the long tetra- and pentam-ethylene bridges lead to an increase in the interarene distances [128]. Therefore the α-carbons of C_3-bridges are oriented *endo* (towards the iron atom) and of C_4- and C_5-bridges *exo* (away from the iron atom) [129]. The conformation of fcp is determined by the number and position of trimethylene bridges (Fig. 37) [129b, c]. The squeezing the iron atom is counteracted with bending of both Cp-rings. Introduction of another, unattached trimethylene group, decreases the bending angle substantially and increases the squeezing of the iron atom compared to ferrocene (70) itself (Fig. 37) [128, 129c,e, 130–134].

4.2 Mössbauer Spectroscopy and Redox Potentials of Ferrocenophanes

Mössbauer spectroscopy and redox potentials play a certain role in fcp chemis-try because the iron to Cp-ring distance is manipulable to a certain extent and therefore the electron density at the iron atom can be varied. Examinations in this series show a linear relationship [135,136] between the iron to ring distance and the Mössbauer parameters—quadrupol splitting and isomer shift [48a]. The same is observed for the redox potentials, which decrease with an increase in Fe-Cp-distances [135, 137]. Consequently Mössbauer parameters and redox potential show a linear relationship [136, 137]. Deviations are observed for fcp containing C_5-bridges and among the other fcp, differences are small, so that the results should be dealt with carefully [137].

4.3 Electron Spectroscopy

Electron spectra of $[4_n]$fcp show a linear relationship of the wave length of the d-d*-absorption band dependent of the number of C_4-bridges [71a, 138]: the d-d*-absorption band is shifted hypsochromic [71a] in going from ferrocene (70) (λ443 nm) to [4]superferrocenophane (77) (λ403 nm). Alkyl substituted fcp as well as fcp containing homoannular bridges, do not show this dependency; consequently slight distortions of the ferrocene nucleus caused by heteroannular bridging, must be responsible for the hypsochromic shift [71a].

In the series of $[3_n]$fcp unusual large hypsochromic shifts of the d-d*-absorption band occur, showing a minimum at $[3_4][4]$fcp (78) (λ345 nm) [72]. The d-d*-absorption band of fcp bearing C_5-bridges is generally shifted to longer wavelengths with a maximum at $[5_4]$ (1, 2, 3, 4)fcp (157) (λ460 nm) [73].

Photoelectron spectroscopy revealed orbital interactions between the π-decks in 8 to be present [50, 139] and in 6 to be absent [34b].

The charge transfer absorption band of the chromium to benzene ring transition in 28 and its derivatives is found at considerably longer wavelengths than in the reference complex 158a (Fig. 38) [29, 40, 123]. The corresponding bis(tricarbonylchromium) complexes show a further bathochromic shift [29, 40,

complex	158a	26	28	35a	36a	32	33	35b	36b
λ_{max} [nm]	318	322	341	342	341	327.5	358	346	342

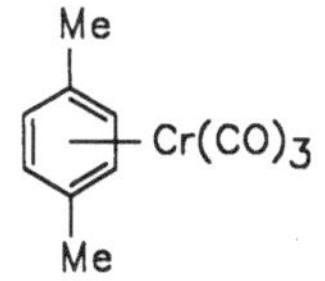

for complexes

26/28/32/33   (see Fig. 5, 6a)

for complexes

35a / 36a / 35b / 36b   (see Fig. 6c)

158a

Fig. 38. Ultraviolet absorption of mono- and bis(tricarbonylchromium) complexes in comparison to an open chained analogous complex (**158a**)

123]. This is due to stabilization of the excited state in the cyclophane ligand compared to the open chain analog. The electron withdrawing effect of a second $Cr(CO)_3$-group leads to a further increase in stability of the excited state; as is seen in the complexes **35b** and **36b**; this effect decreases when the distance of the $Cr(CO)_3$-groups is increased (Fig. 38) [29, 40].

The mono- and bis(tricarbonyl) complexes of **6** (**26** and **32**) absorb at only slightly longer wavelengths than **158a** because of an absent or only a weakly developed π-π-interaction (FIg. 38) [34b, 117].

The ultraviolet spectra of iron-II-complexes of **8** and its derivatives are also different from reference complexes with planar benzene rings [92].

4.4 NMR-Spectroscopic Properties

NMR spectra of cyclophanes are exceptional with respect to open chained benzene analogs (Fig. 39).

The benzene rings in **8** are placed directly on top of one another, resulting in a shielding of the ring protons in ^{1}H-NMR spectra by about $\Delta\delta = 0.58$, because of the ring current of the opposite ring [89].

The π-clouds of *anti*-**6** penetrate one another only partly, so that protons 4-H, 5-H and 6-H (12-H, 13-H and 14-H respectively) absorb in the same range as the arene protons in **158** (Figs. 39, 40), whereas the inner proton 8-H (16-H respectively) are affected by the shielding cone of the opposite benzene ring, resulting in an upfield shift by about $\Delta\delta \approx 3$ (Fig. 40) [35, 89, 140]. Absorption of the arene protons in the *syn*-conformation (*syn*-**6**) is comparable to that in **8**, due to similar conformation (Fig. 39) [41a].

In ^{13}C-NMR spectra, the strained carbons (the quarternary C-3 and C-6 in **8**, and tertiary C-8 and C-16 in *anti*-**6**) absorb at unexpected low field (Figs. 39,

J. Schulz and F. Vögtle

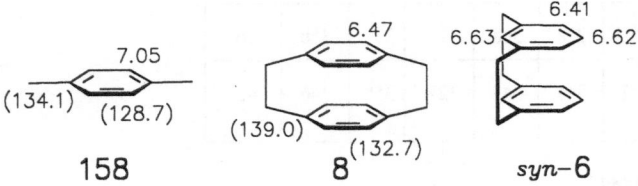

Fig. 39. ^{13}C-NMR (in brackets) and ^{1}H-NMR data of [2.2]cyclophanes in comparison to an open chained analog [142]

40). A decreased electron density at these carbons, due to transannular repulsion, is assumed [141].

Complexation with transition metals leads to a significant upfield shift of the arene protons [16, 35, 89, 142, 143] as well as the ring carbons [89, 142–144]. Examination of proton NMR-spectra of chromium-, iron- and ruthenium complexes particularly with *anti*-6 as ligand provide an insight into the reason for this upfield shift [89, 90].

The chemical shift difference ($\Delta\delta$) of complexes **26** [35], **159** [16] and **160** [89] resemble one another and differ only in the amount of the respective shift (Fig. 40): The protons of the complexed benzene ring are strongly shielded, whereas the ones of the uncomplexed ring are slightly deshielded. Protons 12,13,14-H are shifted downfield by $\Delta\delta = 0.19$ upon complexation with $Ru(C_6Me_6)$ (complex **160**), but the inner proton 16-H by $\Delta\delta = 0.81$ [89]. This additional downfield shift of $\Delta\delta = 0.62$ must be due to a decreased ring current in the complexed benzene ring, whereas the downfield shift of 12,13,14-H must be primarily due to electron withdrawal of the complexed ring [89] (Fig. 40). Other effects like magnetic anisotropy of the metal atom or rehybridization, seem to be less important [89].

The mutual influence of the two benzene rings in *anti*-6 reveals that transannular π-π-interaction cannot be absent completely [89].

Mori [30, 31, 81, 98, 145] pointed out, that the highfield shift of the ligand carbons in complexed [2$_n$]cyclophanes is dependent on their geometry. On the basis of the boatlike distorted conformation of the bridged benzene rings, the distances of the respective ligand carbons to the metal atom vary. In carbon NMR-spectra of iron-[30, 98], chromium-[30, 145] and ruthenium-complexes [81] the complexation shifts ($\Delta\delta$) increase with decreasing distance, independent of the metal (Fig. 41).

In complexes that carry **19** and **8** as ligands, the metal is coordinated on the convex side of the benzene ring. The quarternary bridgehead carbons (C-9, C-10 and C-3, C-6, respectively) have greater distances to the metal atom than the tertiary carbons and consequently are shifted less strongly (Fig. 41). In the corresponding complexes of **6**, the metal is bound to the concave side of the benzene ring. The ligand carbons directed towards the metal atom (C-5, C-8) are more strongly shielded than the other carbons, which lie within the ring plane (Fig. 41). The distance dependence of the complexation shift ($\Delta\delta$) seems to be

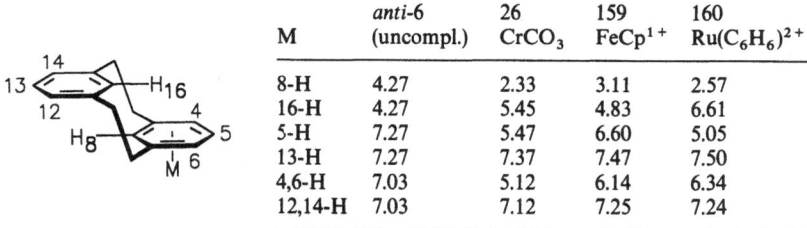

M	anti-6 (uncompl.)	26 CrCO₃	159 FeCp¹⁺	160 Ru(C₆H₆)²⁺
8-H	4.27	2.33	3.11	2.57
16-H	4.27	5.45	4.83	6.61
5-H	7.27	5.47	6.60	5.05
13-H	7.27	7.37	7.47	7.50
4,6-H	7.03	5.12	6.14	6.34
12,14-H	7.03	7.12	7.25	7.24

Fig. 40. Disrupture of the diamagnetic ring current, caused by complexation of transition metals [16, 35, 89] (δ-values in ppm)

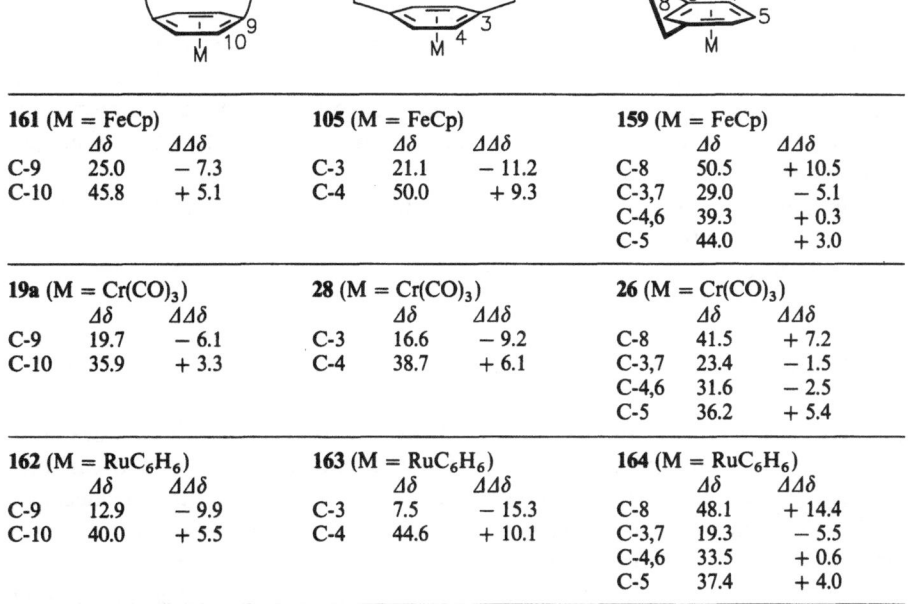

161 (M = FeCp)			105 (M = FeCp)			159 (M = FeCp)		
	$\Delta\delta$	$\Delta\Delta\delta$		$\Delta\delta$	$\Delta\Delta\delta$		$\Delta\delta$	$\Delta\Delta\delta$
C-9	25.0	− 7.3	C-3	21.1	− 11.2	C-8	50.5	+ 10.5
C-10	45.8	+ 5.1	C-4	50.0	+ 9.3	C-3,7	29.0	− 5.1
						C-4,6	39.3	+ 0.3
						C-5	44.0	+ 3.0

19a (M = Cr(CO)₃)			28 (M = Cr(CO)₃)			26 (M = Cr(CO)₃)		
	$\Delta\delta$	$\Delta\Delta\delta$		$\Delta\delta$	$\Delta\Delta\delta$		$\Delta\delta$	$\Delta\Delta\delta$
C-9	19.7	− 6.1	C-3	16.6	− 9.2	C-8	41.5	+ 7.2
C-10	35.9	+ 3.3	C-4	38.7	+ 6.1	C-3,7	23.4	− 1.5
						C-4,6	31.6	− 2.5
						C-5	36.2	+ 5.4

162 (M = RuC₆H₆)			163 (M = RuC₆H₆)			164 (M = RuC₆H₆)		
	$\Delta\delta$	$\Delta\Delta\delta$		$\Delta\delta$	$\Delta\Delta\delta$		$\Delta\delta$	$\Delta\Delta\delta$
C-9	12.9	− 9.9	C-3	7.5	− 15.3	C-8	48.1	+ 14.4
C-10	40.0	+ 5.5	C-4	44.6	+ 10.1	C-3,7	19.3	− 5.5
						C-4,6	33.5	+ 0.6
						C-5	37.4	+ 4.0

Fig. 41. Geometry dependent carbon absorption in ¹³C-NMR spectra; $\Delta\delta = \delta_{cyclophane} - \delta_{cyclophane-complex}$ (complexation shift); $\Delta\Delta\delta = \Delta\delta_{cyclophane} - \Delta\delta_{benzene}$ (conformation shift)

directly correlated with the reciprocal orbital interaction of ligand carbon and metal [146].

Strained cyclophanes as ligands for chromium metal, are ideal objects for investigations of the magnetic environment at the periphery of di(benzene)-chromium (**52**), because of the preservation of their conformation in solution. For the purpose of drawing up a map of the cone of anisotropy around **52**,

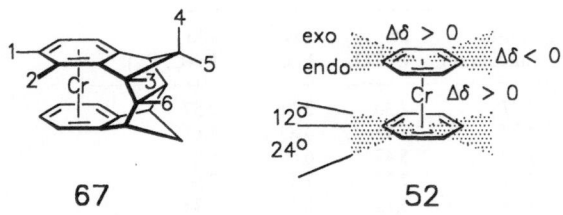

67   **52**

Fig. 42. A model compound (**67**) for determining the magnetic environment of di(benzene)-chromium (**52**) [59]

Elschenbroich prepared several model compounds like **124–126** [39, 108, 147] (Fig. 28) and **67** [59] (Figs. 14, 42). The protons in the bridge of **67** function as monitor protons, because of their rigid position relatively to the di(benzene)chromium unit (Fig. 42). Therefore only proton 3-H, situated almost exactly within the ring plane, is shifted upfield ($\Delta\delta < 0$), whereas the *endo*-proton 6-H (between the two planes) and the *exo*-protons 4-H and 5-H (above the plane) are shifted downfield ($\Delta\delta > 0$) (Fig. 42) [59]. The cone of anisotropy cannot be symmetrical, because both ring currents work in the *endo*-, whereas only one works in the *exo*-region (Fig. 42) [59].

An analogous study of the magnetic environment of ferrocene (**70**) has been known for a long time, due to the higher number of compounds, and led to a similar picture [148].

4.5 Chemical Properties

The redox properties are by far the most extensively examined ones and therefore dealt with separately in Sect. 4.6. The chemical properties of tricarbonylchromium complexes of arenes have been well known for about three decades [123a, 149], however they were transferred to cyclophane chemistry only recently [150]. The outstanding feature of the $Cr(CO)_3$-group is the strong electron withdrawal from the 6π-electron system. This increases the kinetic acidity of arene protons, which can be abstracted with strong bases, and the resulting (η-aryllithium)$Cr(CO)_3$-complexes can be quenched with strong electrophiles [151]. De Meijere et al. [150] obtained substituted paracyclophanes (Fig. 43a) and Vögtle and Schulz [45, 116, 117] prepared chiral metacyclophanes (Fig. 43b) using the reaction sequence.

Derivatization of **26** shows only a low selectivity between the 4- and 5-position. The presence of a dithiane group in each bridge however reveals a clear preference for 5-substitution. Functional group inversion of dithiane groups led to the preparation of numerous other metacyclophanes [116, 117]. Unsymmetrically substituted metacyclophanes are chiral and the corresponding enantiomers were separated [152]. Recently the enantiomeric pure cyclophane **171** was

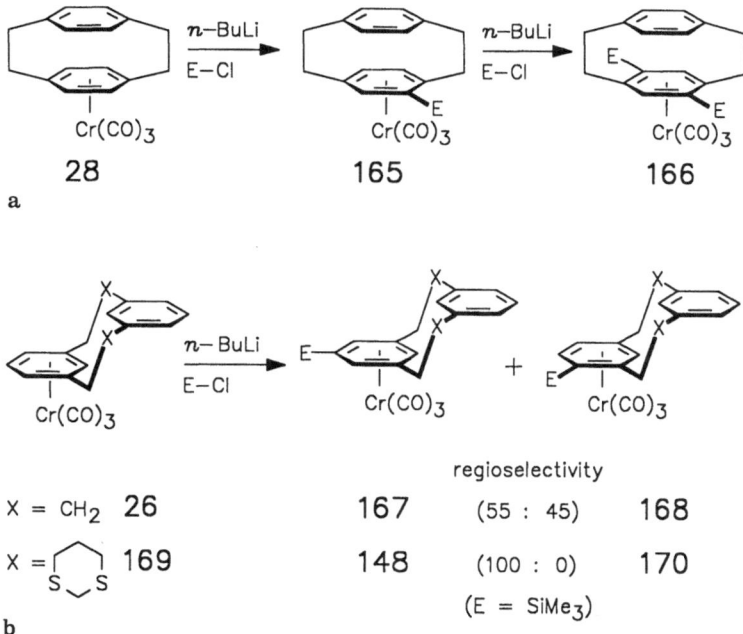

a

b

X = CH₂ 26

X = ⟨S⌣S⟩ 169

regioselectivity

167 (55 : 45) 168

148 (100 : 0) 170

(E = SiMe₃)

Fig. 43a,b. Derivatization of **a**. [2.2]paracyclophane-Cr(CO)₃ [150] and **b**. [2.2]metacyclophane-Cr(CO)₃ complexes [45, 116, 117] via lithiation/electrophilic trapping (E–Cl = Me₃Si–Cl)

complexed with retention of configuration (giving **172**) and finally derivatized to **173** (Fig. 44a) [45]. The absolute configuration of **173** was established by X-ray structure analysis (Fig. 44b) [45]. It interestingly brings together two elements of chirality, namely those of the phane (planar-, helical-chirality) [152, 154] and those of the metallocene [153, 155].

Therefore, the above reaction path provides a convenient way of preparing helical-(planar-) chiral compounds, which show interesting circulardichroisms [45, 153]. These results contribute to a better understanding of the relation between structure and chiroptical properties [153, 154].

Furthermore, attack of electrophiles on [2.2]metacyclophanes easily leads to transannular ring closure yielding dihydropyrenes [156], which points out another advantage of the reaction sequence shown in Figs. 43 and 44.

Certain substituted ferrocenes are also optically active and have been examined regarding this feature [33, 153, 155]. A central challenge for 10–15 years was the complete bridging of ferrocene (**70**), which culminated in the preparation of e.g. "[4]superferrocenophane", which now offers the possibility of including an iron atom in the cage of a saturated hydrocarbon [157, 158]. Unfortunately studies of the hydrogenation of fcp revealed that reduction becomes more difficult when the number of bridges increases [159]. This implies a participation of the iron atom in the reduction step to be probable [159].

J. Schulz and F. Vögtle

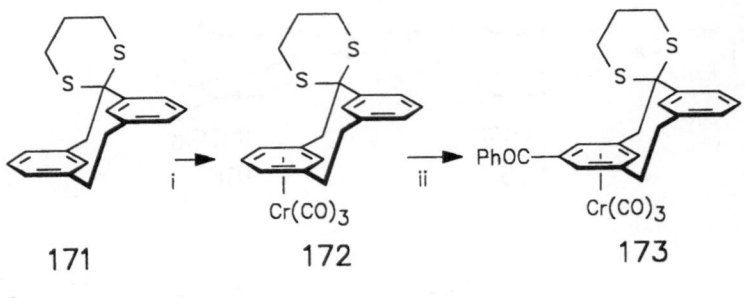

a

b   173

Fig. 44a. Stereoselektive complexation of a chiral [2.2]metacyclophane with Cr(CO)$_3$(NH$_3$)$_3$/THF (i) [45] and regioselective lithiation/electrophilic trapping with PhCOCl (ii); **b.** [45] X-ray structure of complex **173** containing two elements of chirality (plane of chirality in the phane and the "metallocene" chirality)

An interesting but rarely observed reaction is the "*retro*-Friedel-Crafts acylation"; it has been known for years and studied with fcp as substrates [160]. The α-oxo-propylene bridge of **174** rearranges in the presence of Lewis acids to give **175** and **176**; the missing of **177** and **178** in the reaction mixture makes it evident that the Cp-ring-acyl-bond was cleaved exclusively (Fig. 45) [160].

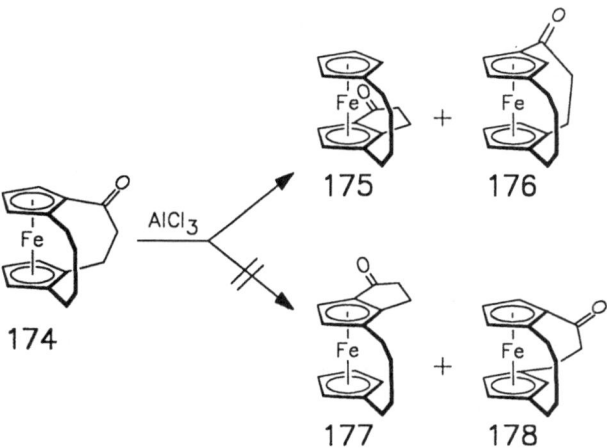

Fig. 45. An example of a *retro*-Friedel-Crafts acylation [160]

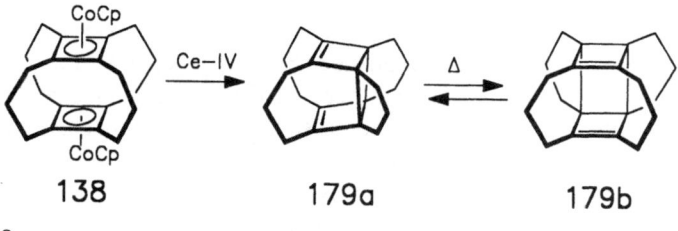

a

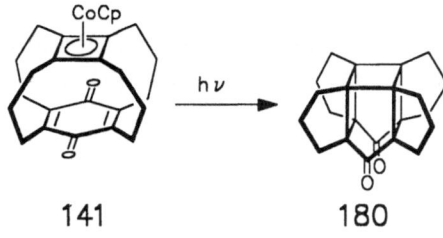

b

Fig. 46. Valence isomerism of cyclophanes containing *anti*-aromatic π-decks after removal of the stabilizing metal [112, 162]

A number of interesting compounds is available, when the coordinated metal atom of cyclophane complexes bearing *anti*-aromatic π-decks is removed, because of the resulting cyclophane is instable and subjected to valence isomerism [161].

J. Schulz and F. Vögtle

In this context, **138** was converted into propella[3₄]prismane (**179**) via an intramolecular cycloaddition reaction, after the stabilizing CpCo-group had been removed photochemically (Fig. 46a) [112, 162].

Similarly, irradiation of the cyclobutadienophane **141** yielded **180** (Fig. 46b) [112].

4.6 Reduction and Oxidation Properties

Of great importance in transition metal chemistry of cyclophanes is that the complexes provide suitable building blocks to construct an organometallic polymer of general structure D (Fig. 3), monomeric and oligomeric units of which are already known (Figs. 25, 27).

To achieve electrical conductivity the metal atoms have to be in different oxidations states [90]. Since the discovery of the Cruetz-Taube complex **181** [164] the delocalization of electrons in various sandwich complexes has been examined [165, 166]. As was shown, the unpaired electron in the bis(fulvalene)bis(iron) ion **182** is completely delocalized and regarded as an ion of class III; the two iron atoms exist in a mixed valence state (Fig 47) [166, 167]. The ion **183** belongs to class II, because a considerable energy barrier in the interaction of the two metal centers was established (Fig. 47) [166, 167]. The ion **184** is not suitable for the design of a polymer, because the unpaired electron is completely localized at one of the chromium atoms, so that one of them carries a positive charge, while the other remains uncharged (Fig. 47) [166, 168].

Cyclophanes are suitable ligands for such mixed valence ions of iron and ruthenium in particular, because of the ease of preparation and its stronger π-bonding ability with arenes [169].

Reduction of bis(hexamethylbenzene)ruthenium (**185**) **with sodium yields 186** [170a], which shows a fluctuating structure [170b]; one arene ligand now is η-coordinated and bent (Fig. 48) [170c]. Cycline voltammetry revealed, that the 2-electron reduction process proceeds in two distinct 1-electron steps [171].

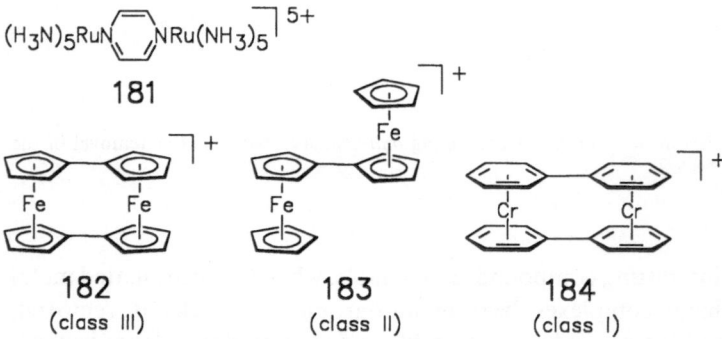

Fig. 47. Classification of mixed valence ions [169]

Exchange of one C_6Me_6-ligand through a cyclophane ligand shows a strong dependence of the redox potential from the phanes geometry [89, 90, 170, 171]. In "superphane" (9) both benzene rings are nearly planar [4a] and can hardly assume the geometry involved with η^4-coordination [170]. Therefore the redox potentials of 185 and 187 are very similar (Fig. 48) [169, 172]. [2_4](1,2,3,4)Cyclophane (189) possess strongly distorted benzene rings [173] and can adopt the conformation necessary for η^4-coordination much better; this consequently results in a large decrease of the potential (Fig. 48) [90].

Complexation of both benzene rings of 192 or 8 give the multidecker sandwich complexes 193 and 195 respectively [174]. Cyclic voltammetry of 193 clearly demonstrates two distinct 1-electron reduction steps [90]. On the basis of its rigid structure [175] η^4-coordination is less easily adopted than in hexamethylbenzene, even though its redox potential is lower [89]. This implies that the phane geometry is widely preserved, and so 194 probably represents the first example of a 20-electron complex of ruthenium (Fig. 49) [90].

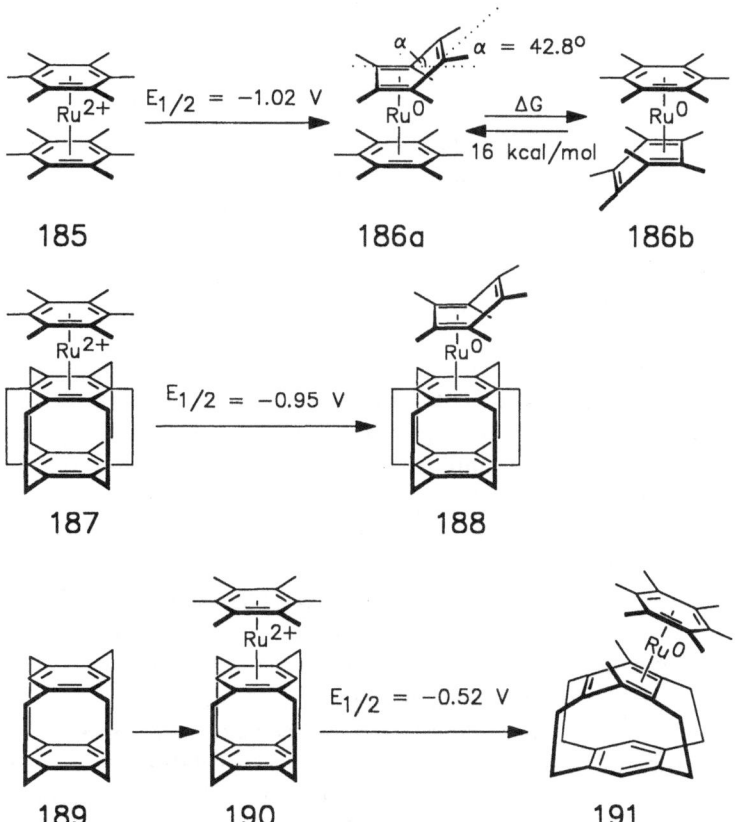

Fig. 48. Redox potentials of bis(η^6-arene)ruthenium complexes as a function of the ligand

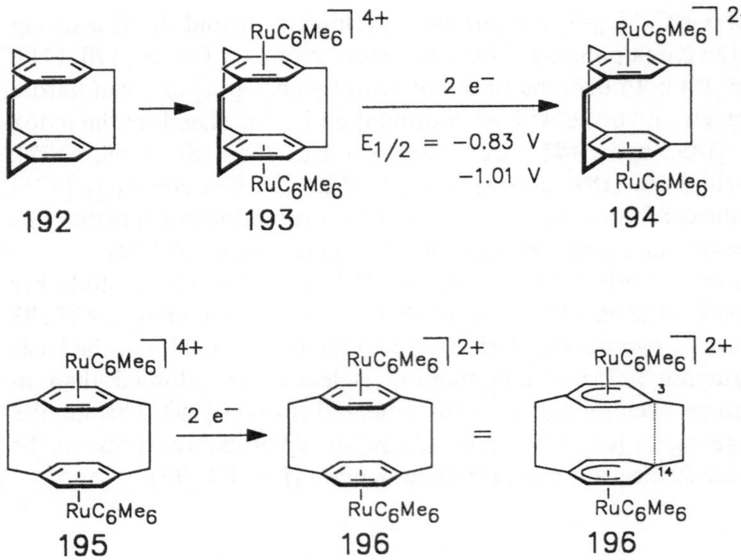

Fig. 49. Geometric consequences for the cyclophane ligand on two electron reduction of its bis(ruthenium) complex [90]

Therefore, cyclophane ligands generally lead to a decrease of redox potentials compared to the hexamethylbenzene ligand, because in the first place cyclophanes change the geometry and in the second place they receive electrons much easier [90].

Structural changes, accompanied with the reduction of ruthenium complexes, were extensively investigated in the case of the 2-electron reduction product of **195** (**196** in Fig. 49) [174]. The X-ray structure analysis of **196** shows an unusually small distance between the quarternary carbons C-3 and C-14; the measured value of 196 pm seems to be the longest carbon single bond determined by X-ray structure analysis till now [174]. The conformation of the two benzene rings is best described as similar to those found in cyclohexadienyl anions. Comparison with the spectroscopic data of other bivalent positive bis(ruthenium) complexes led to the assumption that the cyclohexadienyl anionic π-decks are present as well [174].

Complex **186** can be regarded as an "electron reservoir" compound [176] and is a suitable agent for achieving the reduction of ruthenium complexes. So **197** is reduced to the dication **198** and further on to the uncharged complex **199**, where the single reduction steps are reversible (Fig. 50) [177].

The complex **198** is of particular interest, because the mechanism of electron transfer between two metal centers, which are in different oxidation states, can be studied. NMR-spectra, obtained at room temperature, infer a symmetrical molecule, consisting of a Ru^{1+}-Ru^{1+}-system [177]. At lower temperatures, the signals broaden and split into two groups, inferring two localized π-d-electron

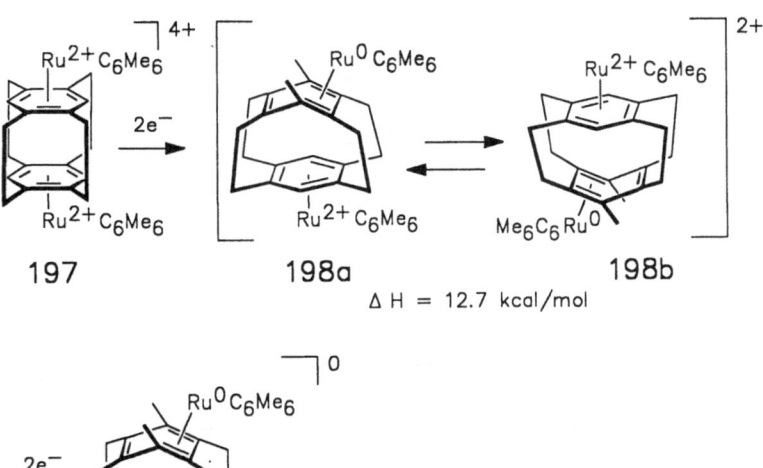

Fig. 50. Geometric consequences of two and four electron reduction of the complex **197** [177]

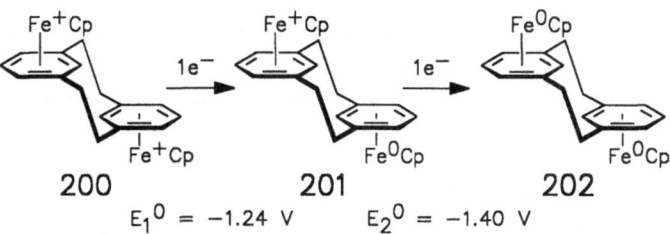

Fig. 51. Transannular interaction in [2.2]metacyclophane: two electron reduction of its bis(iron) complex **200** [180]

systems and two conformations, separated by a measurable energy barrier (Fig. 50) [177]. It is suggested that electron transfer occurs in two 1-electron steps [177]. This establishes the first example of a netto-2-electron intervalence transfer of a discrete, mixed-valence organometallic complex [177], **198** is a class II mixed valence ion and it forms red crystals which have a metallic sheen [169].

Comparisons of the redox properties with those of bis(ruthenium) complexes of polycyclic arenes like phenanthrene [178, 179] show that the electron transfer here can be even faster [179].

The redox properties of complexes of the cobalt triade with [2ₙ]cyclophanes [78] as well as iron [180] were also examined and show a similar behaviour as

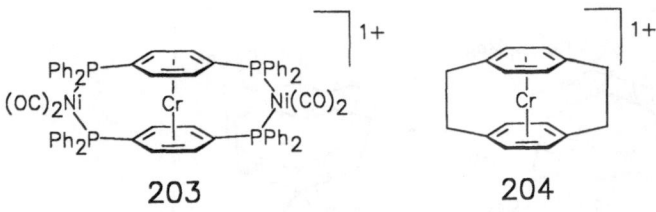

Fig. 52. Radial cations of η^{12}-coordinated chromium atoms [20]

the ruthenium series: Generally, complexes bearing **189** as ligand are always the easiest to reduce. The reduced species are more labile than the corresponding ruthenium complexes. Cobalt-II- forms more stable arene complexes but due to its paramagnetic properties it is unsuitable for electrochemical studies [78]. The change in hapticity, accompanied with the transition from Ru^{2+} to Ru^0 seems to be less important in the iron series [180].

The distance between the iron atom and the cyclophane ligand in the complex **199**, as well as the amount of π-π-transannular interaction determine the potential difference of the first and second reduction step and therefore the amount of interaction between the two iron atoms [180]. A single 2-electron reduction step is expected, when transannular interaction in metacyclophanes is absent, but the cyclic voltammetry clearly shows two 1-electron reduction waves (Fig. 51) [180].

Cyclic voltammetry and photoelectron spectroscopy of $[2_2](1,3)$cyclophane (**6**) itself proved that there is a weak interaction between the two π-decks [181]. Complexes in which a chromium atom is captured within the cavity of a cyclophane (Figs. 11–14, 52) have also been important subjects for studying redox reactions.

In view of a mutual influence of two electron transfer centres within one molecule, **65** was investigated (Fig. 52) [20]. The first oxidation step is reversible and takes place at the chromium atom; the potential is at more positive values than with corresponding open chain complexes [20]. The second oxidation step is irreversible and this infers that the nickel atom has been oxidized, because in contrast the second oxidation in complex **53** is reversible [182].

5 Future Perspectives

Complex chemistry of cyclophanes links the two best examined two-layered molecules of organic and organometallic chemistry, namely $[2_2]$para-cyclophane and ferrocene. The rich redox chemistry of ruthenium, cobalt, rhodium and iridium therefore provides a basis for constructing polymeric structures, of which some oligomeric building units are already known.

Introduction of the tricarbonylchromium group into cyclophane chemistry has great potential for the preparation of planar- (or helical-)chiral compounds. Besides structure/chiroptics relationships, carrying out stereoselective reactions is of major importance. Recently, inclusion of ferrocenes and benzene-$Cr(CO)_3$ into the cavity of a cyclodextrine was achieved and this makes the connection to supramolecular chemistry [183].

Interesting prospects are to be expected in complex chemistry with main group elements as their number is increasing constantly.

Acknowledgements: We thank Dipl. Chem. J. Gross for his helpful ideas and additions and also the Stiftung Volkswagenwerk, which financially supported our work in the field and therefore stimulated the writing of this report.

References

1. Pellegrin M (1899) Rec Trav Chim Pays Bas 18: 558
2. a. Cram DJ, Steinberg HJ (1951) J Am Chem Soc 73: 5691; b. Cram DJ (1959) Rec Chem Prog 20: 71
3. Boekelheide V (1980) Acc Chem Res 13: 65
4. a. Sekine Y, Boekelheide, V (1981) J Am Chem Soc 103: 1777; b. Sekine Y, Brown M, Boekelheide V (1979) J Am Chem Soc 101: 3126
5. a. Jenneskens LW, de Kanter FJJ, Kraakman PA, Turkenburg LAM, Koolhaas WE, de Wolf WH, Bickelhaupt F, Tobe Y, Kakiuchi K, Odaira Y (1985) J Am Chem Soc 107: 3716; b. Rice JE, Lee TJ, Remington RB, Allen WD, Clabo Jr DA, Schäfer III HF (1987) J Am Chem Soc 109: 2902; c. There is evidence for the short life existence of even [3]paracyclophane: F. Bickelhaupt, Amsterdam (personal communication, 1992)
6. a. Otsubo T, Mizogami S, Otsubo I, Tozuka Z, Sakagami A, Sakata Y, Misumi S (1973) Bull Chem Soc Jpn 46: 3519; b. Misumi S, Otsubo T (1978) Acc Chem Res 11: 251
7. Schmidbaur H (1985) Angew Chem 97: 893; Int Ed Engl 24: 893
8. Watts WE (1967) Organomet Chem Rev 2: 231
9. Mueller-Westerhoff UT (1986) Angew Chem 98: 700; Int Ed Engl 25: 702
10. a. Schmidbaur H, Bublak W, Huber B, Hofmann J, Müller G (1989) Chem Ber 122: 265; b. Schmidbaur H, Hager R, Huber B, Müller G (1987) Angew Chem 99: 354; Int Ed Engl 26: 355; c. Schmidbaur H, Bublak W, Huber B, Müller G (1987) Zeitschr Naturforsch B42: 147; d. Schmidbaur H, Bublak W, Huber B, Müller G (1986) Helv Chim Acta 69: 1742; e. Schmidbaur H, Bublak W, Huber B, Müller G (1986) Organomet 5: 1647
11. Probst T, Steigelmann O, Riede J, Schmidbaur H (1990) Angew Chem 102: 1471; Int, Ed Engl 29: 1397
12. Probst T, Steigelmann O, Riede J, Schmidbaur H (1991) Chem Ber 124: 1089
13. Vögtle F, Neumann P (1970) Tetrahedron 26: 5847
14. a. Boekelheide V (1983) Top Curr Chem 133: 87; b. Vögtle F (ed) (1990) Cyclophan-Chemie, Teubner, Stuttgart; (1993) Cyclophane chemistry, Wiley, Chichester; c. Keehn PM, Rosenfeld SM (ed) (1983) Cyclophanes I, II, Academic Press, New York, London
15. Koray AR, Ziegler ML, Blank NE, Haenel MW (1979) Tetrahedron Lett 26: 2465
16. Swann RT, Boekelheide V (1982) J Organomet Chem 231: 143
17. Elschenbroich C, Hurley J, Metz B, Massa W, Baum G (1990) Organomet 9: 889
18. Kasahara A, Izumi T, Yoshida Y, Shimizu I (1982) Bull Chem Soc Jpn 55: 1901
19. Osborne AG, Whiteley RH, Meads RE (1980) J Organomet Chem 193: 345
20. Elschenbroich C, Sebbach J, Metz B, Heikenfeld G (1992) J Organomet Chem 426: 173
21. a. Dorer B, Diebold J, Weyand O, Brintzinger HH (1992) J Organomet Chem 427: 245; b. Lang

H, Seyferth D (1991) Organomet 10: 347; c. Gomez R, Cuenca T, Royo P, Pellinghelli MA, Tiripicchio A (1991) Organomet 10: 1505; d. Burk MJ, Colletti SL, Halterman RL (1991) Organomet 10: 2998; e. Brintzinger HH (1990) In: Dötz KH, Brintzinger RW (eds) Organic synthesis via organometallics, Vieweg, Braunschweig, p 33

22. a. Grossel MC, Goldspink MR, Hriljac JA, Weston SC (1991) Organomet 10: 851; b. Holwerda RA, Robison TW, Bartsch RA, Czech BP (1991) Organomet 10: 2652; c. Rao SJ, Milberg CI, Petter RC (1991) Tetrahedron Lett 32: 3775; d. Beer PD, Kocian O, Mortimer RJ, Spencer P (1992) J Chem Soc Chem Commun 602

23. a. Marr G, Rockett BW (1991) J Organomet Chem 416: 399; b. Silverthorn WE (1975) Adv Organomet Chem 13: 47; c. Pidcock A, Smith JD, Taylor BW (1967) J Chem Soc A 872

24. a. Fischer EO (1957) Angew Chem 69: 715; b. Fischer EO, Öfele K (1957) Chem Ber 90: 2532; c. Fischer EO, Öfele K, Essler H, Fröhlich W, Mortensen JP, Semmlinger W (1958) Chem Ber 91: 2763

25. a. Hudecek M, Toma S (1990) J Organomet Chem 393: 115; b. Schumann H, Arif AM, Richmond TG (1990) Polyhedron 9: 1677; c. Beswick PJ, Greenwood CS, Mowlem TJ, Nechvatal G, Widdowson DA (1988) Tetrahedron 44: 7325; d. Nicholls B, Whiting MC (1959) J Chem Soc 559

26. a. Wey HG, Betz P, Butenschön H (1991) Chem Ber 124: 465; b Tate DP, Knipple WR, Augl JM (1962) Inorg Chem 1: 433; c. Werner H, Deckelmann K, Schönenberger U (1970) Helv Chim Acta 53: 2002; d. Gracey DEF, Jackson WR, Jennings WB, Mitchell TRB (1969) J Chem Soc (B): 1204; e. Müller P, Bernardinelli G, Jacquier Y (1988) Helv Chim Acta 71: 1328

27. a. Kündig EP, Perret C, Spichiger S, Bernardinelli G (1985) J Organomet Chem 286: 183; b. Traylor TG, Stewart KJ (1986) J Am Chem Soc 108: 6977

28. a. Tobe Y, Nakayama A, Kobiro K, Kakiuchi K, Odaira Y (1989) Chem Lett 1549; b. Tobe Y, Ueda K, Kakiuchi K, Odaira Y, Kai Y, Kasai N (1986) Tetrahedron 42: 1851

29. Ohno H, Horita H, Otsubo T, Sakata Y, Misumi S (1977) Tetrahedron Lett 265

30. Mori N, Takamori M (1986) Magnet Res Chem 24: 151

31. Takamori M, Mori N (1986) J Organomet Chem 301: 321

32. a. Cram DJ, Wilkinson DI (1960) J Am Chem Soc 82: 5721; b. Cram DJ, Dewhirst K (1959) J Am Chem Soc 81: 5963

33. Benedikt M, Schlögl K (1978) Monatsh Chem 109: 805

34. a. Kainradl B, Langer E, Lehner H, Schlögl K (1972) Liebigs Ann Chem 766: 16; b. Langer E, Lehner H (1973) Tetrahedron 29: 375

35. Langer E, Lehner H (1979) J Organomet Chem 173: 47

36. Christiani F, De Fillippo D, Deplano P, Devillanova F, Diaz A, Trogu EF, Verani G (1975) Inorg Chim Acta 12: 119

37. Mourad AF, Hopf H (1979) Tetrahedron Lett 1209

38. a. Burri K, Jenny W (1967) Helv Chim Acta 50: 1978; b. Norinder U, Tanner D, Wennerström O (1983) Tetrahedron Lett 24: 5411

39. Elschenbroich C, Schneider J, Wünsch M, Pierre JL, Baret P, Chautemps P (1988) Chem Ber 121: 177

40. Misumi S, Otsubo T (1978) Acc Chem Res 11: 251

41. a. Mitchell RH, Vinod TK, Bushnell GW (1990) J Am Chem Soc 112: 3487; b. Mitchell RH, Vinod TK, Bodwell GJ, Weerawarna KS, Anker W, Williams RV, Bushnell GW (1986) Pure Appl Chem 58: 15; c. Mitchell RH, Vinod TK, Bushnel GW (1985) J Am Chem Soc 107: 3340

42. a. Böckmann K, Vögtle F (1981) Chem Ber 114: 1065; b. Kamp D, Boekelheide V (1978) J Org Chem 43: 3470

43. a. Balbach BK, Koray AR, Okur A, Wülknitz P, Ziegler ML (1981) J Organomet Chem 212: 77; b. Koray AR, Ziegler ML (1980) J Organomet Chem 202: 13; c. Koray AR, Ziegler ML (1979) J Organomet Chem 169: C34

44. Vögtle F, Przybilla KJ, Mannschreck A, Pustet N, Büllesbach P, Reuter H, Puff H (1988) Chem Ber 121: 823

45. Schulz J, Bartram S, Nieger M, Vögtle F (1992) Chem Ber 125: 2553

46. Zaworotko MJ, Stamps RJ, Ledet MT, Zhang H, Atwood JL (1985) Organomet 4: 1697

47. Adams CM, Holt EM (1990) Organomet 9: 980

48. a. Elschenbroich C, Salzer A (1989) Organometallics VCH, Weinheim; b. Watts L, Fitzpatrick JD, Pettit R (1965) J Am Chem Soc 87: 3253; c. Efraty A (1977) Chem Rev 77: 691; d. Pettit R (1975) J Organomet Chem 100: 205

49. Jutzi P, Siemeling U, Müller A, Bögge H (1989) Organomet 8: 1744

50. a. Heilbronner E, Yang Z-z (1984) Top Curr Chem 115: 1; b. Gerson F (1984) Top Curr Chem 115: 57
51. a. Rajasekharan MV, Giezynski S, Ammeter JH, Oswald N, Michaud P, Hamon JR, Astruc D (1982) J Am Chem Soc 104: 2400; b. Thompson MR, Day CS, Day VW, Mink RI, Mutterties EL (1980) J Am Chem Soc 102: 2979
52. a. Timms PL (1981) In: Müller A, Diemann E (eds) Transition metal chemistry, Verlag Chemie, Weinheim, p 23; b. Kündig EP, Timms PL (1980) J Chem Soc Dalton 991; c. Timms PL (1969) J Chem Soc Chem Commun 1033
53. Elschenbroich C, Möckel R, Zennek U (1978) Angew Chem 90: 560; (1978) Angew Chem, Int Ed Engl 17: 531
54. Hope H, Bernstein J, Trueblood KN (1972) Acta Cryst Sect B28: 1733
55. a. Haaland A (1965) Acta Chem Scand 19: 41; b. Keulen E, Jellinek F (1966) J Organomet Chem 5: 490; c. Förster E, Albrecht G, Dürselen W, Kurras E (1969) J Organomet Chem 19: 215
56. a. Haenel MW, Flatow A (1979) Chem Ber 112: 249; b. Gantzel P, Trueblood KN (1965) Acta Cryst 18: 958
57. Sakurai H, Hoshi S, Kamiya A, Hosomi A, Kabuto C (1986) Chem Lett 1781
58. Elschenbroich C, Hurley J, Massa W, Baum G (1988) Angew Chem 100: 727; (1988) Angew Chem, Int Ed Engl 27: 684
59. Elschenbroich C, Schneider J, Prinzbach H, Fessner WD (1986) Organomet 5: 2091
60. Pierre JL, Baret P, Chautemps P, Armand M (1981) J Am Chem Soc 103: 2986
61. Cohen-Addad PC, Baret P, Chautemps P, Pierre JL (1983) Acta Cryst C39: 1346
62. Kang HC, Hanson AW, Eaton B, Boekelheide V (1985) J Am Chem Soc 107: 1979
63. Griffith EAH, Amma EL (1974) J Am Chem Soc 96: 743 and 5407
64. Kealy TJ, Pausen PL (1951) Nature 168: 1039
65. Rockett BW, Marr G (1991) J Organomet Chem 416: 327
66. a. Rinehart KL, Frerichs AK, Kittle PA, Westmann LF, Gustafson DH, Pruett RL, McMahon JE (1960) J Am Chem Soc 82: 4111; b. Pruett RL, Morehouse EL, Patent US (1962) 3,063,974; (1963) Chem Abstr 58: 11404
67. Hisatome M, Watanabe N, Sakamoto T, Yamakawa K (1977) J Organomet Chem 125: 79
68. Hisatome M, Kawajiri Y, Yamakawa K (1982) J Organomet Chem 226: 71
69. Lüttringhaus A, Kullich W (1958) Angew Chem 70: 438
70. Hisatome M, Kawajiri Y, Yamakawa K, Harada Y, Iitaka Y (1982) Tetrahedron Lett 23: 1713
71. a. Hisatome M, Watanabe J, Kawajiri Y, Yamakawa K, Iitaka Y (1990) Organomet 9: 497; b. Hisatome M, Watanabe J, Yamakawa K, Iitaka Y (1986) J Am Chem Soc 108: 1333
72. Watanabe J, Hisatome M, Yamakawa K (1987) Tetrahedron Lett 28: 1427
73. Hisatome M, Yamashita R, Watanabe J, Yamakawa K, Iitaka Y (1988) Bull Chem Soc Jpn 61: 1687
74. Hisatome M, Kawajiri Y, Yamakawa K, Mamiya K, Harada Y, Iitaka Y (1982) Inorg Chem 21: 1345
75. a. Astruc D, Batail P, Martin ML (1977) J Organomet Chem 133: 77; b. Batail P, Grandjean D, Astruc D, Dabard R (1976) J Organomet Chem 110: 91; c. Astruc D, Dabard R, Martin M, Batail P, Grandjean D (1976) Tetrahedron Lett 829
76. Jutzi P (1986) Chem Rev 86: 983
77. Jutzi P, Krallmann R, Wolf G, Neumann B, Stammler HG (1991) Chem Ber 124: 2391
78. Plitzko KD, Boekelheide V (1988) Organomet 7: 1573
79. Elzinga B, Rosenblum M (1982) Tetrahedron Lett 23: 1535
80. a. Bennett MA, Mathesen TW (1979) J Organomet Chem 175: 87; b. Bennett MA, Mathesen TW, Roberts GB, Smith AK, Tucker PA (1980) Inorg Chem 19: 1014; c. Domaille PJ, Ittel SD, Jesson JP, Sweigart DA (1980) J Organomet Chem 202: 191
81. Miura T, Horishita T, Mori N (1987) Organomet J Chem 333: 387
82. a. Swann RT, Hansen AW, Boekelheide V (1984) J Am Chem Soc 106: 818; b. Swann RT, Boekelheide V (1984) Tetrahedron Lett 25: 899
83. Rohrbach WD, Boekelheide V (1983) J Org Chem 48: 3673
84. Koelle U, Fuss B, Rajasekharan MV, Ramakrishna BL, Ammeter JH, Böhm MC (1984) J Am Chem Soc 106: 4152
85. Thompson SJ, Bailey PM, White C, Maitlis PM (1976) Angew Chem 88: 506; (1976) Angew Chem, Int Ed Engl 15: 490
86. Rybinskaya MI, Kudinov AR, Kaganovich VS (1983) J Organomet Chem 246: 279

J. Schulz and F. Vögtle

87. Gill TP, Mann KR (1982) Organomet 1: 485
88. Laganis ED, Finke RG, Boekelheide V (1980) Tetrahedron Lett 21: 4405
89. Laganis ED, Voegeli RH, Swann RT, Finke RG, Hopf H, Boekelheide V (1982) Organomet 1: 1415
90. Swann RT, Hansen AW, Boekelheide V (1986) J Am Chem Soc 108: 3324
91. Cram DJ, Helgesen RC, Lock D, Singer LA (1966) J Am Chem Soc 88: 1324
92. Elzinga J, Rosenblum M (1983) Organomet 2: 1214
93. Nesmeyanov AN, Vol'kenau NA, Shiloutseva LS (1970) Dokl Akad Nauk SSSR 190: 857
94. a. Schirsch PFT, Boekelheide V (1981) J Am Chem Soc 103: 6873; b. Gill TP, Mann KR (1980) Inorg Chem 19: 3007
95. Gill TP, Mann KR (1981) J Organomet Chem 216: 65
96. Laganis ED, Finke RG, Boekelheide V (1981) Proc Natl Acad Sci USA 78: 2657
97. Koray AR (1981) J Organomet Chem 212: 233
98. Mori N, Takamori M (1985) J Chem Soc Dalton Trans (1985) 1661
99. Hiermeier J, Köhler FH, Müller G (1991) Organomet 10: 1787
100. Siemeling U, Jutzi P (1992) Chem Ber 125: 31
101. Atzkern H, Hiermeier J, Köhler FH, Steck A (1991) J Organomet Chem 408: 281
102. a. Atzkern H, Hiermeier J, Kanellakopulos B, Köhler FH, Müller G, Steigelmann O (1991) J Chem Soc Chem Commun 997; b. Miller JS, Calabrese JC, Epstein AJ, Bigelow RW, Zhang JH, Reiff WM (1986) J Chem Soc Chem Commun 1026; c. Broderick WE, Thompson JA, Day EP, Hoffmann BM (1990) Science 249: 401
103. a. Hopf H, Raulfs FW, Schomburg D (1986) Tetrahedron 42: 1655; b. Hopf H, Raulfs FW (1985) Israel J Chem 25: 210
104. a. Wilkinson G (1956) Org Synth 36: 31; b. Eich JJ, King RB (1965) Organomet Synth 1: 73; c. King RB, Bisnette MB (1964) Inorg Chem 3: 796
105. a. Hopf H, Dannheim J (1988) Angew Chem 100: 724; Int Ed Engl 27: 701; b. El-Tamany S, Raulfs FW, Hopf H (1983) Angew Chem 95: 631; Int Ed Engl 22: 633
106. Frim R, Raulfs FW, Hopf H, Rabinovitz M (1986) Angew Chem 98: 160; (1986) Angew Chem, Int Ed Engl 25: 174
107. Elschenbroich C, Spangenberg B, Mellinghoff H (1984) Chem Ber 117: 3165
108. Elschenbroich C, Schneider J, Mellinghoff H (1987) J Organomet Chem 333: 37
109. a. Schmidbaur H, Bublak W, Haenel MW, Huber B, Müller G, (1988) Z Naturforsch 43b: 702; b. Schmidbaur H, Bublak W, Huber B, Reber G, Müller G (1986) Angew Chem 98: 1108; (1986) Angew Chem, Int Ed Engl 25: 1089; c. Kawashima N, Kawashima T, Otsubo T, Misumi S (1978) Tetrahedron Lett 5025
110. Streitwieser Jr A, Müller-Westerhoff U (1968) J Am Chem Soc 90: 7364
111. Garbe JE, Boekelheide V (1983) J Am Chem Soc 105: 7384
112. Gleiter R (1992) Angew Chem 104: 29; (1992) Angew Chem Int Ed Engl 31: 27
113. Gleiter R, Karcher M, Ziegler ML, Nuber B (1987) Tetrahedron Lett 28: 195
114. Gleiter R, Treptow B, Kratz D, Nuber B (1992) Tetrahedron Lett 33: 1733
115. a. Brown CJ (1953) J Chem Soc 3278; b. Kai Y, Yasuoka N, Kasai N (1977) Acta Crystallogr B33: 754
116. a. Vögtle F, Schulz J, Nieger M (1991) Chem Ber 124: 1415; b. Schulz J, Nieger M, Vögtle F (1992) J Chem Soc Perkin Trans II, in press
117. Schulz J, Nieger M, Vögtle F (1991) Chem Ber 124: 2797
118. a. Koray AR, Zahn T, Ziegler ML (1985) J Organomet Chem 291: 53; b. Koray AR (1982) J Organomet Chem 232: 345
119. Hanson AW (1982) Cryst Struct Commun 11: 1019
120. Hanson AW (1962) Acta Cryst 15: 956
121. Cameron TS, Linden A, Sturge KC, Zaworotko MJ (1992) Helv Chim Acta 75: 294
122. Hunter AD, Shilliday L, Furey WS, Zaworotko MJ (1992) Organomet 11: 1550
123. de Meijere A, Reiser O, Stöbbe M, Kopf J, Adiwidjaja G, Sinnwell V, Kahn SI (1988) Acta Chem Scand A42: 611
124. Kai Y, Yasuoka N, Kasai N (1978) Acta Cryst B34: 2840
125. a. Benn R, Blank NE, Haenel MW, Klein J, Koray AR, Weidenhammer K, Ziegler ML (1980) Angew Chem 92: 45; (1980) Angew Chem, Int Ed Engl 19: 44; b. Blank NE, Haenel MW, Koray AR, Weidenhammer K, Ziegler ML (1980) Acta Cryst B36: 2054
126. Oike H, Kai Y, Miki K, Tanaka N, Kasai N (1987) Bull Chem Soc Jpn 60: 1993

127. a. Hanson AW (1982) Cryst Struct Commun 11: 901; b. Hanson AW (1982) Cryst Struct Commun 11: 1395
128. Hillman M, Fujita E, Dauplaise H, Kvick A, Kerber RC (1984) Organomet 3: 1170
129. a. Hisatome M, Yamashita R, Watanabe J, Yamakawa K (1988) Bull Chem Soc 61: 1391; b. Hisatome M, Watanabe J, Yamakawa K, Kozawa K, Uchida T (1984) J Organomet Chem 262: 365; c. Hisatome M, Kawajiri Y, Yamakawa K, Kozawa K, Uchida T (1982) J Organomet Chem 236: 359; d. Hisatome M, Hillman M (1981) J Organomet Chem 212: 217; e. Spaulding LD, Hillmann M, Williams GJB (1978) J Organomet Chem 155: 109
130. Laing MB, Trueblood KN (1965) Acta Cryst 19: 373
131. a. Jones ND, Marsh RE, Richards JH (1965) Acta Cryst 19: 330; b. Batail P, Grandjean D, Astruc D, Dabard R (1975) J Organomet Chem 102: 79
132. Paul IC (1966) J Chem Soc Chem Commun 377
133. Hillman M, Fujita E (1978) J Organomet Chem 155: 87
134. a. Seiler P, Dunitz JD (1979) Acta Cryst B35: 1068; b. Takusagawa F, Koetzle TF (1979) Acta Cryst B35: 1074
135. a. Hillman M, Nagy AG (1980) J Organomet Chem 184: 433; b. Nagy AG, Dezsi I, Hillman M (1976) J Organomet Chem 117: 55
136. Hillman M, Gordon B, Weiss AJ, Guzikowski AP (1978) J Organomet Chem 155: 77
137. Fujita E, Gordon B, Hillman M, Nagy AG (1981) J Organomet Chem 218: 105
138. Hisatome M, Watanabe J, Yamakawa K, Kozawa K, Uchida T (1985) Nippon Kagaku Kaishi 572
139. a. Heilbronner E, Maier JP (1974) Helv Chim Acta 57: 151; b. Boschi R, Schmidt W (1973) Angew Chem 85: 408; (1973) Angew Chem, Int Ed Engl 12: 402
140. Glotzmann C, Langer E, Lehner H (1974) Monatsh Chem 105: 354
141. a. Takemura T, Sato T (1976) Can J Chem 54: 3412; b. Sato T, Takemura T (1974) J Chem Soc Chem Commun 97
142. Mitchell RH, Vinod TK, Bodwell GJ, Bushnell GW (1989) J Org Chem 54: 5871
143. a. Elschenbroich C, Schneider J, Burdorf H (1990) J Organomet Chem 391: 195; b. Maricq MM, Waugh JS, Fletcher JL, McGlinchley MJ (1978) J Am Chem Soc 100: 6902
144. Steele BR, Sutherland RG, Lee CC (1981) J Chem Soc Dalton Trans 529
145. Mori N, Takamori M, Takemura T (1985) J Chem Soc Dalton Trans (1985) 1065
146. Ono I, Mita S, Kondo S, Mori N (1989) J Organomet Chem 367: 81
147. Elschenbroich C, Koch J, Schneider J, Spangenberg B, Schiess P (1986) J Organomet Chem 317: 41
148. Turbitt TD, Watts WE (1972) Tetrahedron 28: 1227
149. a. Solladié-Cavallo A (1985) Polyhedron 4: 901; b. Hegedus LS (1990) J Organomet Chem 380: 169; c. Kündig EP, Desobry V, Simmons DP, Wenger E (1989) J Am Chem Soc 111: 1804; d. Dickins PJ, Gilday JP, Negri JT, Widdowson DA (1990) Pure Appl Chem 62: 575; e. Kündig EP (1985) Pure Appl Chem 57: 1855; f. Kündig EP, Do Thi NP, Paglia P, Simmons DP, Spichiger S, Wenger E (1987) Organometallics in Organic Synthesis (ed de Meijere A, tom Dieck H), Springer
150. Stöbbe R, Reiser O, Thiemann T, Daniels RG, de Meijere A (1986) Tetrahedron Lett 27: 2353
151. a. Uemura M, Nishikawa N, Take K, Ohnishi M, Hirotsu K, Higuchi T, Hayashi Y (1983) J Org Chem 48: 2349; b. Card RJ, Trahanovsky WS (1980) J Org Chem 45: 2560; c. Card RJ, Trahanovsky WS (1980) J Org Chem 45: 2555
152. a. Schlögl K (1984) Top Curr Chem 125: 27; b. Langer E, Lehner H, Schlögl K (1973) Tetrahedron 29: 2473
153. Schlögl K (1989) Organometallics in Organic Synthesis 2 (ed Werner H, Erker G) Springer-Verlag
154. a. Langer E, Lehner H, Schlögl K (1973) Tetrahedron 29: 2473; b. Falk H, Reich-Rohrwig P, Schlögl K (1970) 26: 511
155. a. Schlögl K (1986) J Organomet Chem 300: 219; b. Hirschmann H (1983) Top Stereochem 14: 189; c. Schlögl K (1967) Top Stereochem 1: 39
156. a. Vögtle F, Neumann P (1973) Synthesis 85; b. Akabori S, Sato T, Hata K (1968) J Org Chem 33: 3277; c. Allinger NL, Gorden BJ, Hu S, Ford RA (1967) J Org Chem 32: 2272; d. Mitchell RH, Chaudhary M, Kamada T, Slowey PD, Williams RV (1986) Tetrahedron 42: 1741
157. vgl. Chelat-Komplexe: a. Creaser II, Harrowfield JM, Herlt AJ, Sargeson AM, Springborg J,

Geue RJ, Snow MR (1977) J Am Chem Soc 99: 3181; b. Creaser II, Geue RJ, Harrowfield JM, Herlt AJ, Sargeson AM, Snow MR, Springborg J (1982) J Am Chem Soc 104: 6016

158. vgl. Carceranden: a. Cram DJ, Karbach S, Kim YH, Baczynskyj L, Kalleymeyn GW (1985) J Am Chem Soc 107: 2575; b. Cram DJ (1992) Nature 356: 29

159. Hillman M, Gordon B, Dudek N, Fajer R, Fujita E, Gaffney J, Jones P, Weiss AJ, Takagi S (1980) J Organomet Chem 194: 229

160. Hisatome M, Kawajiri Y, Watanabe J, Masahiro Y, Yamakawa K (1984) J Organomet Chem 266: 147

161. a. Bickelhaupt F, de Wolf WH (1988) Recl Trav Chim Pays-Bas 107: 459; b. Paquette LA, Kesselmayer MA, Underiner GE, House SD, Rogers RD, Meerholz K, Heinze J (1992) J Am Chem Soc 114: 2644

162. Gleiter R, Karcher M (1988) Angew Chem 100: 851; (1988) Angew Chem, Int Ed Engl 27: 840

163. a. Fagan PJ, Ward MD, Calabrese JC (1989) J Am Chem Soc 111: 1698; b. Ward MD, Fagan PJ, Calabrese JC, Johnson DC (1989) 111: 1719; c. Miller JS, Epstein AJ, Reiff WM (1988) Science 240: 40

164. a. Fürholz U, Bürgi HB, Wagner FE, Stebler A, Ammeter JH, Krausz E, Clark RJH, Stead MJ, Ludi A (1984) J Am Chem Soc 106: 121; b. Creutz C, Taube H (1973) J Am Chem Soc 95: 1086; c. Creutz C, Taube H (1969) J Am Chem Soc 91: 3988

165. Elschenbroich C, Bilger E, Metz B (1991) Organomet 10: 2823

166. a. Dong TY, Chou CY (1990) J Chem Soc Chem Commun 1332; b. Dong TY, Hendrickson DA, Pierpont CG, Moore MF (1986) J Am Chem Soc 108: 963; c. Elschenbroich C, Heck J (1979) J Am Chem Soc 101: 6773

167. Le Vanda C, Bechgaard K, Cowan DO, Mueller-Westerhoff UT, Eilbracht P, Candela GA, Collins RL (1976) J Am Chem Soc 98: 3181

168. Elschenbroich C, Heck J (1979) J Am Chem Soc 101: 6773

169. Boekelheide V (1986) Pure Appl Chem 58: 1

170. a. Fischer EO, Elschenbroich C (1970) Chem Ber 103: 162; b. Darensbourg MY, Mutterties EL (1979) J Am Chem Soc 100: 7425; c. Huttner G, Lange S (1972) Acta Cryst B28: 2049

171. Pierce DT, Geiger WE (1989) J Am Chem Soc 111: 7636

172. Finke RG,, Voegeli RH, Laganis ED, Boekelheide V (1983) Organomet 2: 347

173. Hanson AW (1977) Acta Cryst B33: 2003

174. Plitzko KD, Rapko B, Gollas B, Wehrle G, Weakley T, Pierce DT, Geiger Jr WE, Haddon RC, Boekelheide V (1990) J Am Chem Soc 112: 6545

175. Hanson AW (1980) Cryst Struct Commun 9: 1243

176. a. Lacoste M, Rabaa H, Astruc D, Ardoin N, Varret F, Saillard JY, Le Beuze A (1990) J Am Chem Soc 112: 9548; b. Astruc D (1986) Acc Chem Res 19: 377; c. Hamon JR, Astruc D, Michaud P (1981) J Am Chem Soc 103: 758; d. Astruc D, Hamon JR, Althoff G, Roman E, Batail P, Michaud P, Mariot JP, Varret F, Cozak D (1979) J Am Chem Soc 101: 5445

177. Voegeli RH, Kang HC, Finke RG, Boekelheide V (1986) J Am Chem Soc 108: 7010

178. Plitzko KD, Wehrle G, Gollas B, Rapko B, Dannheim J, Boekelheide V (1990) J Am Chem Soc 112: 6556

179. Plitzko KD, Boekelheide V (1987) Angew Chem 99: 715; (1987) Angew Chem, Int Ed Engl 26: 700

180. Bowyer WJ, Geiger WE, Boekelheide V (1984) Organomet 3: 1079

181. Sato T, Torizuka K, Komaki R, Atobe H (1980) J Chem Soc Perkin II 561

182. Elschenbroich C, Sebbach J, Metz B (1991) Helv Chim Acta 74: 1718

183. Stoddart JF, Zarzycki R (1988) Recl Trav Chim Pays-Bas 107: 515

Syntheses and Ionophoric Properties of Crownophanes

Seiichi Inokuma, Shigefumi Sakai, and Jun Nishimura

Department of Chemistry, Gunma University, Kiryu 376, Japan

Table of Contents

Topics in Current Chemistry, Vol. 172
© Springer-Verlag Berlin Hiedelberg 1994

Crown compounds having cyclophane-skeletons and called "crownophanes", are reviewed. They form characteristic host-guest complexes due to their crown-cyclophane hybrid structures. Significant stabilization of complexes is often reported through the ion-dipole, cation-aromatic π-electron and $\pi-\pi$ stacking interactions, size-and-shape complementarity, and hydrophobic interaction.

In order to produce crownophanes having intriguing complexing behavior, some displacements have been used at the cyclization step, although they are not always simple and convenient. Alternatively, a highly efficient intramolecular [2 + 2] photocycloaddition of styrene derivatives was applied to the synthesis of novel crownophanes possessing functional groups on the aromatic nuclei and/or polyether moieties, which show selective complexing abilities with not only valuable metal cations such as Li^+, Ag^+, and Pb^{2+}, but also some organic compounds. Their structural effects on ion-transport and/or extraction abilities are described in detail.

1 Introduction

The name of Crownophane was coined for compounds having both crown ether and cyclophane moieties. It is well known that the term Crown Ether was introduced by Pedersen, when he serendipitously discovered the first examples of this compound class. They are defined as species possessing cyclic polyether structures. Cyclophanes, sometimes simply called Phanes – the name was coined by Cram – are defined as compounds having layered aromatic nuclei or having bridges across the surface of an aromatic nucleus.

Why do we call them by this name? It is because they are hybrids of both structural features, and consequently not only show the ionophoric properties of crown ethers but also the characteristic behavior of cyclophanes, especially the structural advantage to orient several moieties in desired directions. These interesting hybrids steadily growing in number, have thus been given this specific family name which may not be confused with any other families in host-guest chemistry.

In this review a great many examples of crownophanes are reported, which were produced by various kinds of displacement reactions. They are often called by individual names such as calixarene, spherand, etc. Then, we will summarize crownophanes obtained by [2 + 2] photocycloaddition of styrene moieties.

In 1987, Nishimura, one of the authors, and his co-workers discovered that the [2 + 2] photocycloaddition of styrene derivatives proceeds smoothly, when pertinent conditions are carefully chosen. By this method, many cyclophanes including orthocyclophanes, metacyclophanes, paracyclophanes, (4,4')-biphenylophanes, (1,4)-, (1,5)-, and (2,6)naphthalenophanes, (1,6)- and (3,6)phenanthrenophanes, and so forth have been synthesized [1–18]. This method is regarded as one of the most convenient synthetic methods to cyclophanes. In 1988, Inokuma, also one of the authors, joined the Nishimura group in Gunma University. He had built up his career in the area of host-guest chemistry, studying some cooperative carrier systems composed of conventional crown ethers and alkanoic acids [19–21]. So, as mentioned in Sect. 3, two different experiences merged and created an excellent cyclization method for crownophanes. Since then, the research group has successfully developed the preparative method in several ways and has successfully prepared crowno-phanes which extract and transport Li^+-, Ag^+-, or Pb^{2+}-ions selectively, and also double-looped crownophanes which transport size-complementary dibasic acids selectively.

2 Crownophanes Cyclized by Displacement Reactions

There are a great variety of examples in the crownophane family. All these compounds were made by using displacement reactions. This review deals mostly with their properties, but does not refer to their detailed synthetic methods. In this section, compounds are classified by the type of aromatic nuclei, and at the same time, it is emphasized how they work in the host-guest chemistry, taking advantage of the cyclophane skeleton.

2.1 Crownophanes Having Cyclophane Skeletons

Misumi and co-workers made paracyclophane-based crownophanes (1–4) (Structure 1) and concluded that the cyclophane moieties serve as a π-donor for the complexation of alkali metal cations [22].

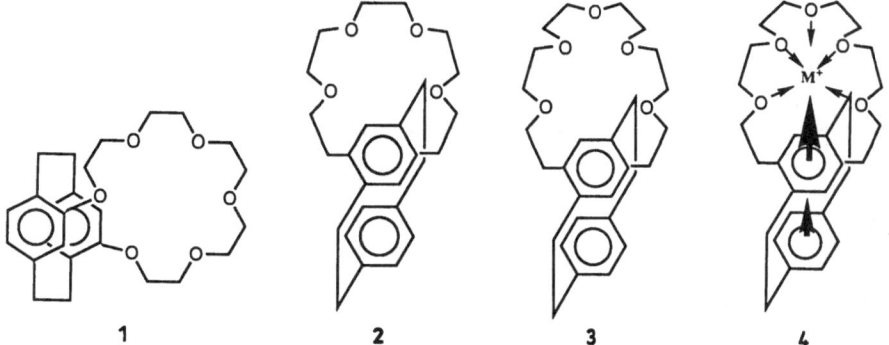

1   **2**   **3**   **4**

Structure 1.

Cram and co-workers described complexation of [2.2]paracyclophane or para-phenylene unit containing crown compounds with primary ammonium salts, diammonium salts, and alkali metal salts [23, 24]. Compound 5 makes complexes with *tert*-butylammonium thiocyanate (1:2 stoichiometry) and hexa-methylenediammonium or decamethylenediammonium hexafluorophosphate (1:1 stoichiometry). Cyclophanes 5–7 (Structure 2) solubilized 2 molar equival-ent of *tert*-butylammonium tetraphenylborate in $CHCl_3$ due to the resulting lipophilic complexes.

Stoddard and his co-workers reported that 34-crown-10 **8** forms complexes with pyridinium salts, [DQT]$^{2+}$ [PF$_6$]$_{-2}$ and [PQT]$^{2+}$ [PF$_6$]$_{-2}$ (both 1:1 stoichiometry) (Structure 3), in which the central parts of the organic cations are placed within the shielding regions of benzene rings; namely the planar π-electron-deficient dications are sandwiched face-to-face between the two π-electron-rich benzene rings [25–27].

Structure 2.

More preorganized macrocycle **9** (Structure 4) forms more stable complexes with [DQT]$^{2+}$ than that of the corresponding dibenzocrown ether [28].

DQT^{2+} PQT^{2+} **Structure 3.**

Bis-metaphenylene-32-crown-10 (**10**) (Structure 4) also forms complexes with the cations mentioned above [29].

Vögtle and his co-workers have synthesized bis-paraphenylenetetraoxo-38-crown-10 (**11**) (Structure 5) in 20% yield by using a high dilution technique and found its complex formation with NaSCN in 1:2 stoichiometry [30].

9   **10**   Structure 4.

11   Structure 5.

Whitlock and his co-workers have prepared tetraoxaparacyclophane **12** (Structure 6) and a tetraoxanaphthalenophane containing four acetylene units as rigid linkages, which have wide-open cavities enclosed by the two separated aromatic rings, and are modified by some functional groups such as carboxylate groups. These functional groups make the crownophanes highly soluble in water. Their complexing abilities with aromatic and aliphatic ionic guests in organic or aqueous phase were reported in detail [31–34].

12   Structure 6.

(D_3)-Bis(cyclotriveratrylene) derivatives [35] as new macrocage hosts (**13**) (Structure 7) possess roughly spherical and considerably rigid lipophilic cavities. They have strong complexing ability toward neutral guests, namely poly-halomethanes, in water [36] and even in lipophilic media [37–39]. Stability

13   **Structure 7.**

constants of the respective complexes are so high that solid complexes have even been isolated [40, 41].

Receptor **14** [42] (Structure 8) was prepared so as to bind a bifunctional guest such as sodium *p*-nitrophenoxide. In the complex, face-to-face π–π, lipophilic, and ion-dipole interactions are considered to work simultaneously. A more rigid three-dimensional receptor analog **15** [43] (Structure 8) was also made. It shows complexing abilities for both potassium *p*-nitrophenoxide and CH_2Cl_2.

14   **15**

Structure 8.

The first member of a new class of crowned calix[4]arenes, **16a** [44] (Structure 9), possesses two cavities. A lipophilic one enclosed by four aromatic nuclei and a hydrophilic one formed by the oligo(oxyethylene) ring and two ionizable phenolic hydroxyl groups. This ionophore binds K^+, NH_4^+, and Ba^{2+} ions.

Other crowned calix[4]arene analogs **16b** and **16c** [44–48] (Structure 9) having various oligo(oxyethylene) chains as bridging and binding parts and substituents at the phenoxyl groups were synthesized. Their ionophoric properties for alkali metal cations and the stoichiometry of the complexes were studied in organic media and solid state. In $CDCl_3$ solution, metal-free compound **16a** exists as a "cone" conformer, whereas it takes a "partial cone" conformation in the presence of potassium picrate and has shown the selectivity value K^+/Na^+ $= 2.7 \times 10^3$ [47]. The highest selectivity value (1.18×10^4) among synthetic

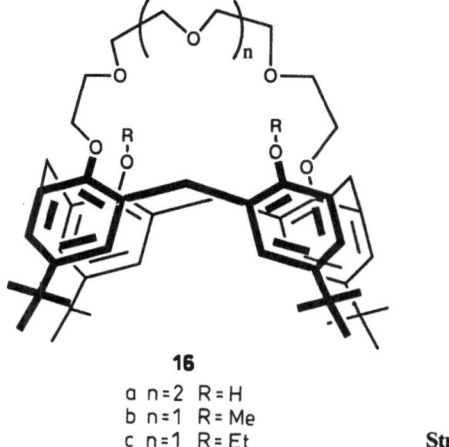

16

a n=2 R=H
b n=1 R=Me
c n=1 R=Et

Structure 9.

ionophores has been observed for the fixed "partial cone" conformer of the ethyl analog **16c** [48].

Calixarene derivatives possessing one and two crown units on the lower rim were readily prepared via a new regioselective 1,2-dealkylation technique of tetraalkylated calixarene using $TiBr_4$. Conformationally rigid cone-formed biscrown-type calix[4]arene **17** (Structure 10) showed selective extractability in the order of $Rb^+ \approx K^+ \gg Na^+ > Cs^+$. This selectivity is believed to be due to its sandwich-type complexation [49].

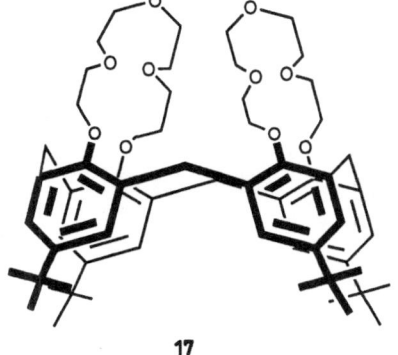

17 **Structure 10.**

Upper rim-modified calixcrowns **18** [50] (Structure 11), which have various oligo(oxyethylene) chains, were prepared as a mixture of two conformers; i.e., the partial cone (P) and the cone (C). The ratio P/C decreased from 4.0 to 1.7 with the increase of the ether chain length and finally reached at the plateau a value of 1.7. By using these compounds (**18a** and **18b**), the conformational interconversion was quantitatively examined by 2-D EXSY NMR spectroscopy.

In order to build better hosts which have 'preorganized' binding sites, Cram and co-workers have prepared "hemispherands" [50] and "spherands" [51],

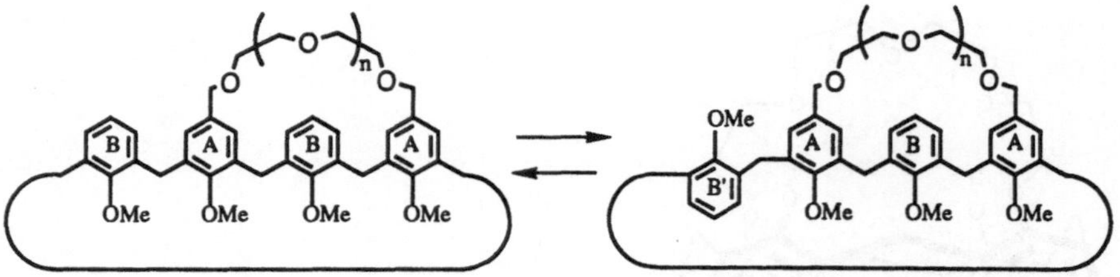

18

a n = 0
b n = 1
c n = 2
d n = 3
e n = 4

Structure 11.

whose unique structures were partly replaced by oligo(oxyethylene) chains to make them to crownophanes. Spherand **19** [51] (Structure 12) having six anisole moieties is a superior host for the small alkali metal cations, like Li$^+$ and Na$^+$, due to the fully organized and rigid structure [52–54]. The spherand not only rejects the larger alkali metal cations such as K$^+$, Rb$^+$, and Cs$^+$, but also shields the cavity from solvent molecules by the anisyl groups. This results in a decrease in the rate of decomposition compared with other flexible crown ethers.

Cram and co-workers [55] also designed a variety of spherands and hemispherands like **20**, and cryptaspherands like **21** based on the rigid prototype **19** (Structure 12).

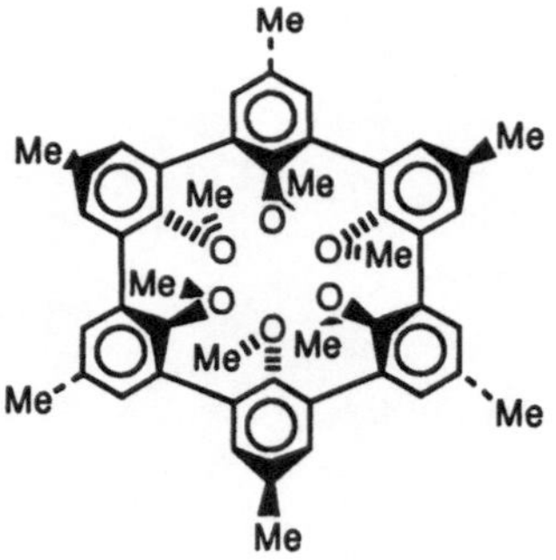

19

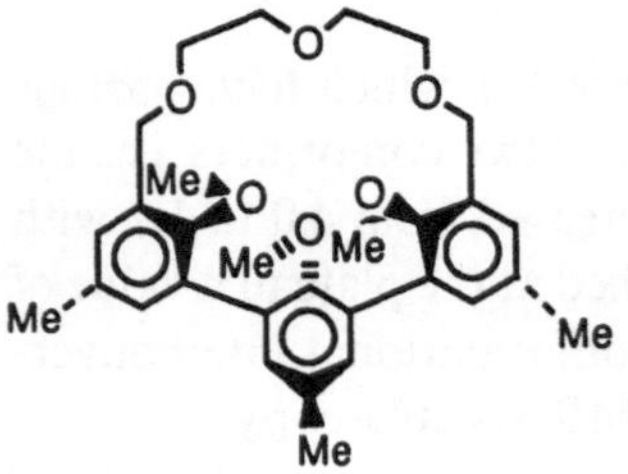

20

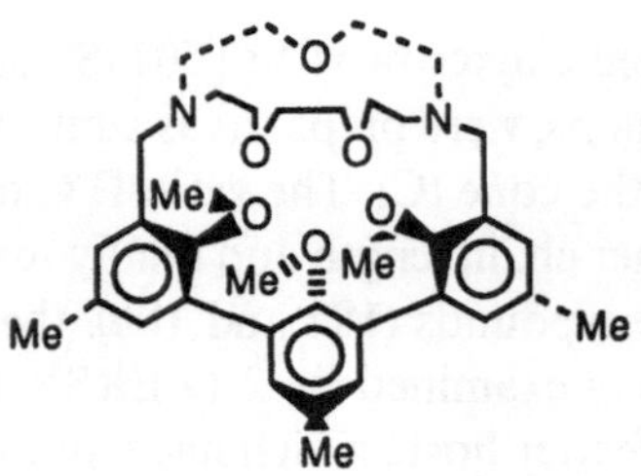

21

Structure 12.

Hemispherands represented by **20** [50], are at least half preorganized for binding metal ions. By the effects of hemispherand structures on complexing properties toward alkali and ammonium picrates in CHCl$_3$, it was found that anisole units become better complexing sites for cations than conventional crown ethers, because the oxygen atoms of anisole units are held in sterically enforced and desired conformations. In these cases, however, the *m*-teranisole units in macroring systems are required to have such conformations. Hemispherands **22** with four self-organized anisole units [56] and the related compounds **23** and **24** (Structure 13) containing one [57] or more [56, 58] cyclic urea units were also prepared and their binding properties were thoroughly examined.

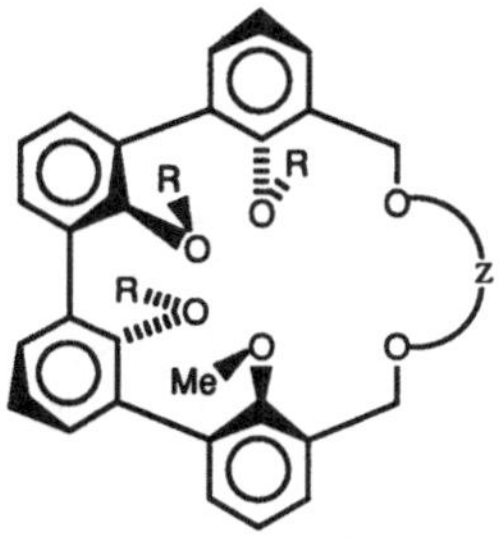

22

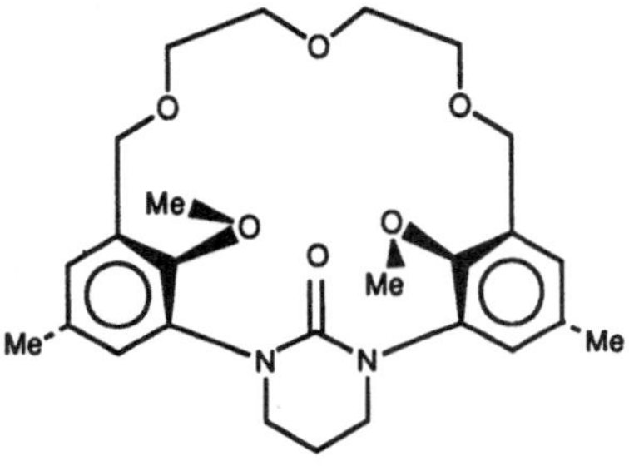

23 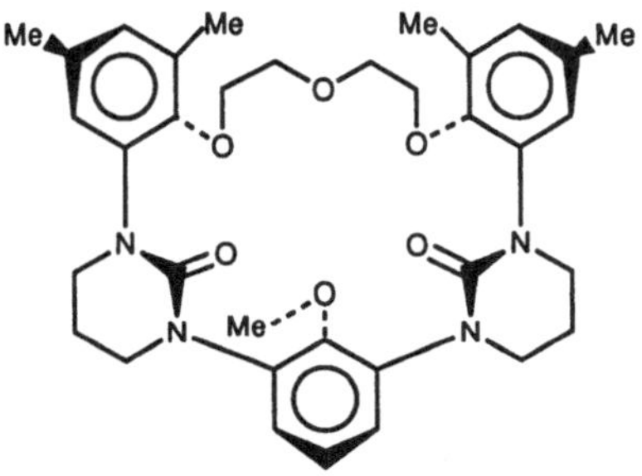   **24**

Structure 13.

Reinhoudt and co-workers reported that hemispherands **25** (Structure 14) substituted by a 4*H*-pyranyl unit [59] show high affinities toward larger alkali metal salts such as Na$^+$ and K$^+$ picrates in CDCl$_3$ system saturated with D$_2$O. This preference is attributed to the lack of the inward pointing methoxyl groups. In general, nitroaryl-containing hemispherand **26** [60, 61] and pyridyl-possessing analogs **27** [60, 62] (Structure 15) exhibit lower complexing abilities for alkali metal cations than Cram's parent hemispherand **20**.

Reinhoudt and co-workers hybridized calixarene and spherand structural features to make a new host **28** [63] (Structure 16) which forms kinetically stable complexes with NaBr and KBr. Because of this high stability, the decomposition

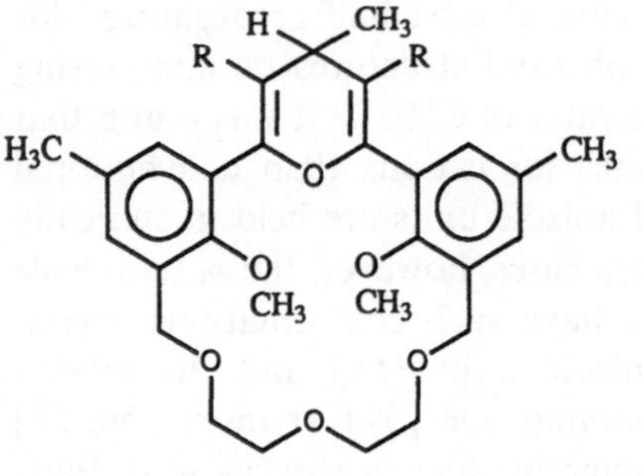

25

Structure 14.

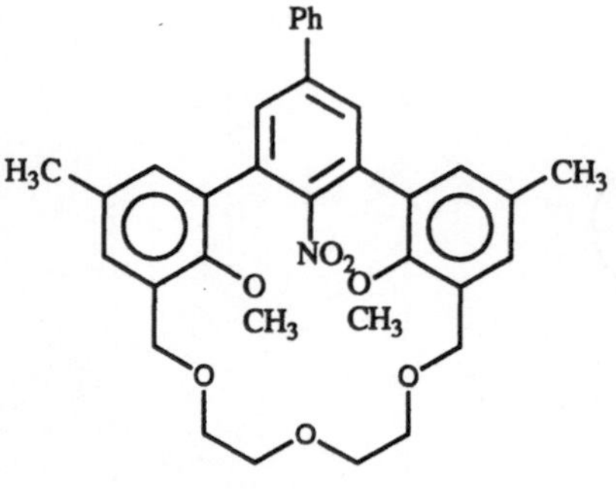

26

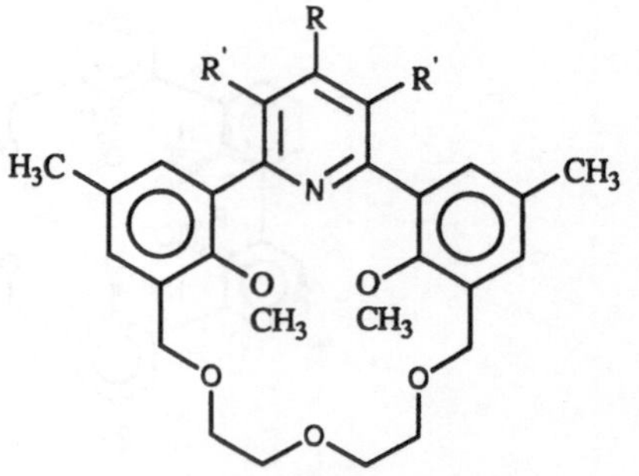

Structure 15.

27

a R=Ph,R'=H
b R=H ,R'=H
c R=H ,R'=Me

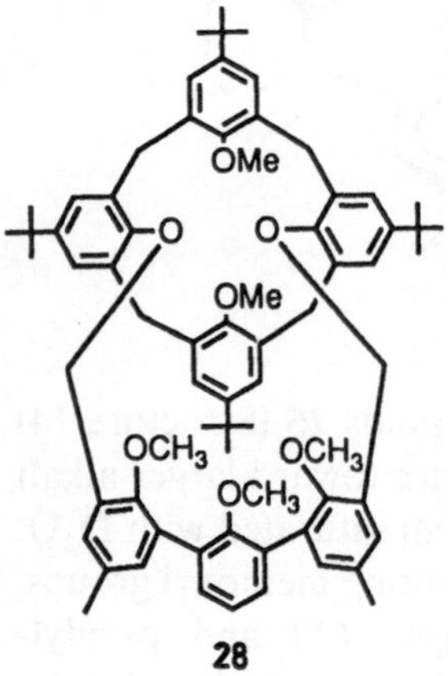

28   Structure 16.

of these complexes in H_2O/CH_3OH (4:1) requires long reaction times (approx. 3 days) and high temperature 120 °C. The high degree of preorganization and the high hydrophobicity of this host prevent solvent molecules from assisting in the decomposition of the complexes.

2.2 Crownophanes Having Naphthalenophane Skeletons

Naphthalenophanes **29a–c** (Structure 17) were reported to act as receptors for organic dications $[PQT]^{2+}$ and $[DQT]^{2+}$ (Structure 3) [64].

29

a n=2, m=1
b n=2, m=2
c n=3, m=3

Structure 17.

Although **29a** and **29b** form 1:1 complexes with these dications in acetone, receptor **29c** complexes with $[PQT]^{2+}$ in a different 1:2 stoichiometry in the solid state [65]. In these cases, electrostatic and charge transfer interactions were recognized as playing significant roles for the supramolecular binding.

2.3 Crownophanes Having Layered Large Aromatic Nuclei

Anthracenocrown species **30** [66, 67], **31** [68], and **32** [69, 70] (Structure 18) of the crownophane family, have been designed in order to control their light emission and photochemistry when binding Na^+ ions in the crown ether rings. Under irradiation, **30** gave different photoproducts in the presence or absence of $NaClO_4$ by reflecting its complexed or uncomplexed species.

Macrotricyclic anthracenyl receptor **31** (Structure 19) having both hydrophilic and hydrophobic parts displays a dual fluorescence emission spectrum (monomer and excimer type emissions). Rubidium perchlorate or 1,7-diaminoheptane dihydrochloride is selectively bound by this receptor. The bindings are considered to occur with only the hydrophilic part or with the simultaneously working hydrophilic and hydrophobic parts. The complexation strongly affected the spectrum [68].

Based on the specific affinities to divalent cations, which are often observed for naturally occurring or synthetic macrolides [69], anthraceno maclolides **32** [70] (Structure 18) have been designed for the optical detection of alkaline earth metal cations. Macrolides **32** also display dual fluorescence, whose behavior depend on the number of n and temperature. Moderate addition effects of $Sr(ClO_4)_2$ or $Ba(ClO_4)_2$ on the changes in UV absorption or fluorescence emission spectra were recognized.

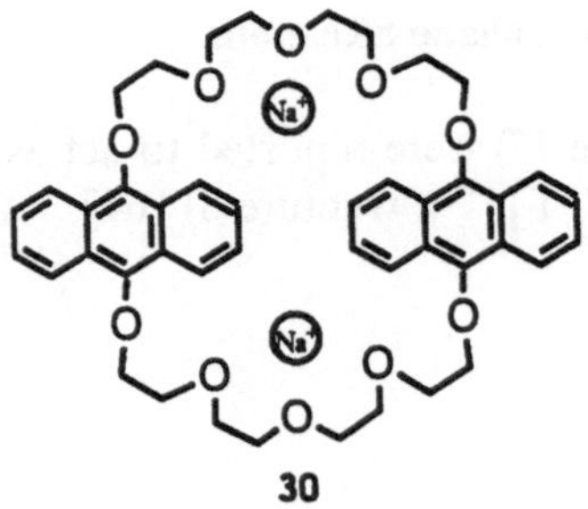

30

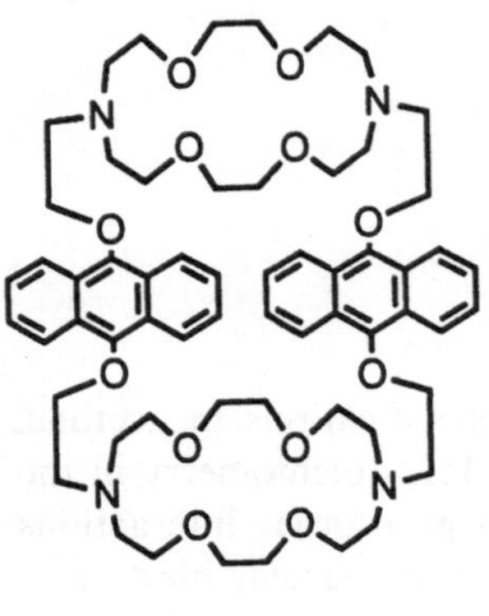

31

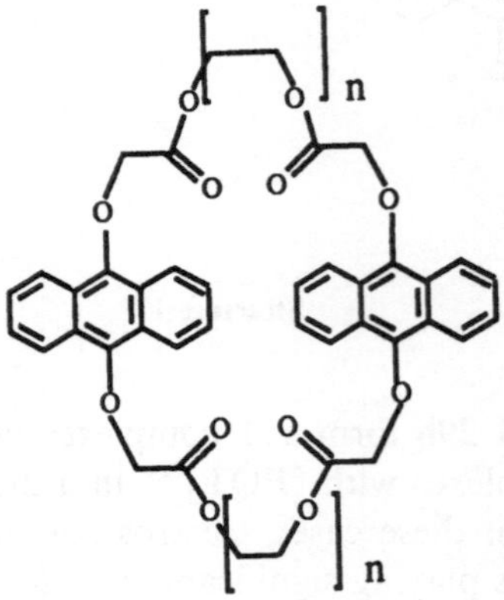

32

a n = 1
b n = 2
c n = 3
d n = 4

Structure 18.

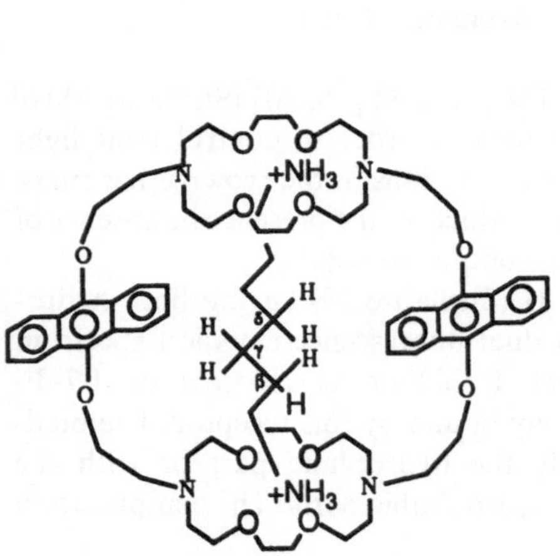

31–H₃N⁺(CH₂)NH₃⁺

$31-H_3N^+(CH_2)NH_3^+$

Structure 19.

As mentioned above, the hybrids of crown ethers and cyclophanes give many opportunities to design selective ionophores, because of the characteristic rigidity of cyclophanes. This design, however, is almost always accompanied by somewhat tedious synthetic tasks. It is the major drawback in this molecular architecture.

3 Crownophanes Cyclized by [2+2] Photocycloaddition Reactions

3.1 Synthesis of Crownophanes

A remarkable difference between these previously reported host molecules and crownophanes presented here is the simplicity and convenience of syntheses. Although there are a few convenient methods for crown ether synthesis such as Okahara's "one-pot" method [71], generally speaking, most of these methods are accompanied by linear byproducts and therefore, give poor yields. As mentioned in the introduction, we found an excellent synthetic method for a new kind of crownophane by means of intramolecular [2 + 2] photocycloaddition of styrene derivatives having long oligo(oxyethylene) linkages [72] (Scheme 1). This photocyclization can easily be applied to prepare some analogs containing heteroatoms other than oxygen in the polyether ring and also that contain some functional groups like hydroxyl groups on the ring. Prototypical crownophanes are synthesized as described below.

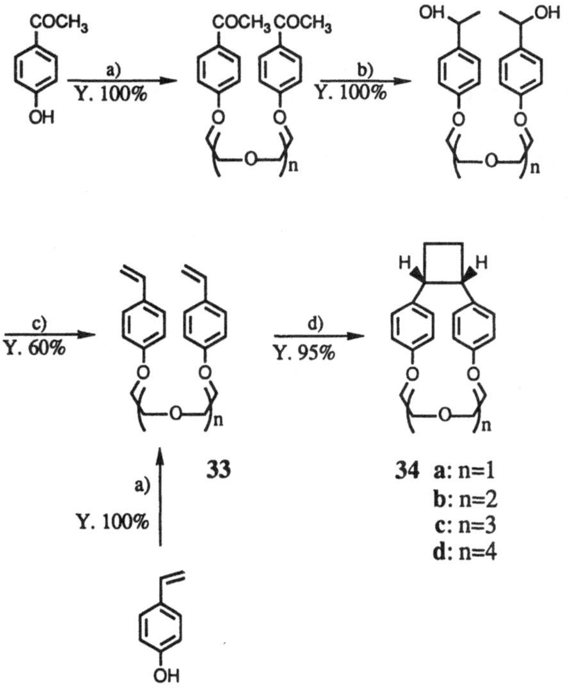

a) TsO(C$_2$H$_4$O)$_{n+1}$Ts, NaOH/dioxane. b) NaBH$_4$/EtOH. c) TsO$^-$HPy$^+$/benzene.
d) hν (> 280 nm)/solv. (MBF$_4$), N$_2$.

Scheme 1.

The [2 + 2] photocycloadditions of styrene derivatives [3] and succinic acid [73] were reported to be affected by the number of the methylene units (the length of the linkage, n); i.e., the yields increase with increasing the number n, reach to the optimum at around $n = 4$, and abruptly decrease from $n = 6$. According to the result, the designed precursors having long oligo(oxyethylene) linkages did not appear to be suitable for these kinds of photoreactions. From the optimistic and serendipitous points of view, however, the photoreaction of the precursors were carried out with and without using the plausible template effects, which are often employed for the syntheses of crown ethers; i.e., the photoreaction was performed in various solvents in the presence or absence of an alkali metal fluoroborate. Contrary to our pessimistic expectations, the reaction proceeded more smoothly than we have ever experienced (Table 1).

The original method beginning from acetylphenols (cf. Scheme 1) has been improved by using commercially available p-vinylphenol and oligo(ethylene glycol) ditosylates in the presence of alkali metal hydroxide. The improved method usually affords corresponding styrene derivatives in nearly quantitative yields.

As shown in Table 1, the crownophanes were obtained in excellent yields. For the synthesis of **34c** (s. Scheme 1), a significant template effect was detected (entries 1 to 3 vs. 4). Figure 1 shows the photoreaction profiles of **33c** (s. Scheme 1) in the absence and presence of $NaBF_4$ in methanol. In both cases (s. Figs. 1a and b) the maximum yields were obtained at the same time when the conversions reached 100% and then the yields decreased with longer irradiation. The photodecomposition of the crownophane was retarded by adding an alkali salt due to the increased stabilization of the crownophane-M^+ complex.

Styrene derivative possessing an oligo(oxyethylene) unit with six ethereal oxygens affords the corresponding crownophane in a nearly quantitative yield

Table 1. Preparation of crownophanes **34**[a]

Entry	Olefin[b]	Reaction conditions			Conv.[e] %	Yield (%)[e]
		Solv.	Add.[c]	Time (min)[d]		**34**
1	**33c**	MeOH	$LiBF_4$	30	100	82
2	**33c**	MeOH	$NaBF_4$	40	100	86
3	**33c**	MeOH	KBF_4	50	100	88
4	**33c**	MeOH	–	20	98	67
5	**33c**	MeCN	–	30	100	91
6	**33c**	PhH	–	30	100	74
7	**33d**	MeOH	$LiBF_4$	20	100	94
8	**33d**	MeOH	$NaBF_4$	20	100	93
9	**33d**	MeOH	KBF_4	30	100	91
10	**33d**	MeOH	–	30	100	93
11	**33d**	MeCN	–	30	100	95

[a] A 400 W high-pressure mercury lamp was set at a distance of 5 cm from Pyrex test tubes (15 ml) which contained the reaction mixture (10 ml) under a nitrogen atmosphere at r.t. [b] 2 mM. [c] 40 mM. [d] At around maximum yields. [e] Determined by GLC, using tetraethylene glycol bis-p-ethylphenyl ether as an internal standard.

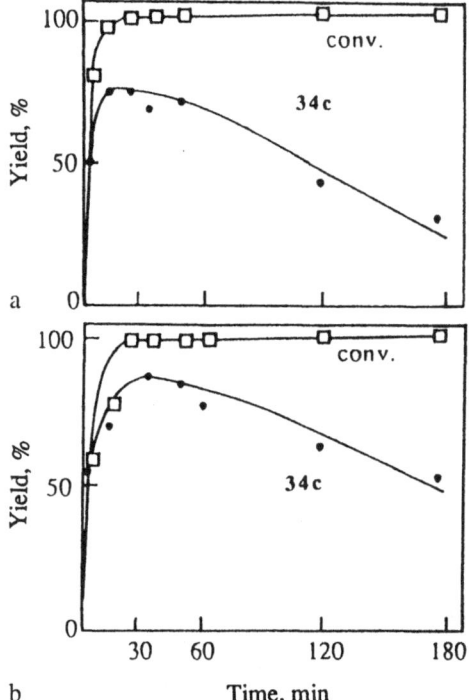

Fig. 1. Time course for the preparation of crownophanes: a In MeOH. b In MeOH with NaBF$_4$

even in the absence of alkali metal salts in various solvents. The high efficiency, despite the long linkages, probably is attributable to the hydrophobic intra-molecular interaction of aromatic moieties in polar solvents and to the dispersion force between the segments of the oligo(oxyethylene) chain in apolar solvents, which reduces its conformational freedom. This is an intriguing result, and the true origin of this phenomenon is unknown.

By the photoreaction of corresponding styrene derivatives, *m*- and *o*-isomers also proceeded in good yields. The large scale photoreactions could be performed by using a concentrated acetonitrile solution (0.1 M) of the precursors. Furthermore, nearly quantitative yields were achieved with irradiation by sunlight through a Pyrex glass filter using an aqueous solution of precursors in the presence of sodium dodecanoate as a surfactant forming micelles to dissolve them in water. This result is important from a technical point of view.

3.2 Complexing Ability of the Crownophanes toward Alkali Metal Cations

In order to examine the ionophoric activities of crownophanes, their extraction of solid alkali metal thiocyanates into CH$_2$Cl$_2$ was carried out. As shown in Table 2, crownophanes **34** (s. Scheme 1) efficiently and selectively extracted LiSCN under anhydrous conditions. The linear oligo(oxyethylene) compounds also extracted the salt sufficiently. Their alkali metal affinity seems to result in

Table 2. Solubilization of thiocyanate salts by ethers into methylene chloride[a]

Ether	Amount of solubilized thiocyanates[b] Molar ratio MSCN/ether		
	KSCN	NaSCN	LiSCN
34c	0.1	0.3	1
33c	0.1	0.4	1
34d	0.3	0.3	1
33d	0.1	0.1	1

[a] Experimental conditions: [polyether], 0.25 mmol/5 ml of methylene chloride; MSCN, 7.5 mmol; stirred at r.t. for 24 h.
[b] From nitrogen content in the organic layer, which was determined by the elemental analysis.

the template effect on the photocyclization to afford the crownophanes in high yields.

3.3 Li⁺-Selective Transport by Crownophanes

The transport of alkali and alkaline earth metal cations by many crown ethers and their derivatives has been extensively studied. Much effort has been paid to increase their selectivity as well as their efficiency. Making highly Li^+ selective ionophores is a primary concern in this field, because large quantities of lithium could be extracted from sea water for use to nuclear fusion generators [74]. Moreover, some carriers of Li^+ selective electrodes to monitor the cation have attracted much attention in the therapy of manic depressive psychosis [75].

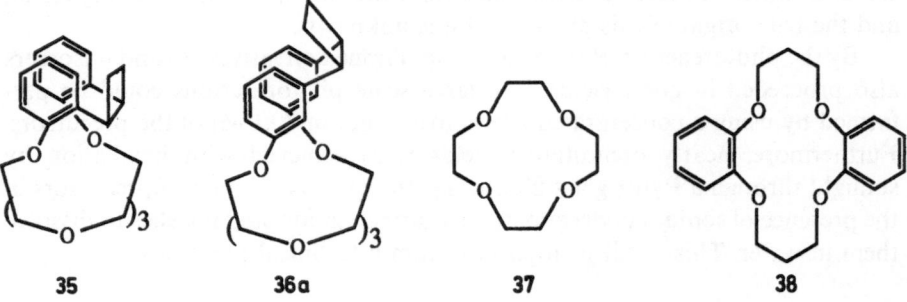

35   **36a**   **37**   **38**

Structure 20.

The most effective and simple transport is a proton-driven one for alkali and alkaline earth metal cations, when the carrier possesses a proton-exchangable functional group such as carboxyl [76], phenolic hydroxyl [77] or amino group [78]. Inokuma, one of the authors, developed a unique proton-driven system for alkali and alkaline earth metal cations by using a mixture of a simple and conventional crown ether or acyclic polyether and an alkanoic acid as a cooperative carrier [19–21]. So we were prompted to apply this cooperative system to the crownophanes and to examine how selectively the transport goes under the conditions shown in Fig. 2.

Data shown in Table 3 with ortho- and metacrownophanes **35** and **36a** (Structure 20) indicate that all crownophanes exhibit Li^+ selectivity toward Na^+ and K^+, and especially metacrownophane **36** shows excellent Li^+ selectivity.

Interestingly the highest selectivity toward Li^+ cations among paracrownophanes is obtained by **34c** (s. Scheme 1) with five ethereal oxygen atoms, whose structure differs from the so-called lithiophilic crown compounds with four ethereal oxygen atoms such as 12-crown-4 (**37**), 14-crown-4, and their derivatives (**38**) (Structure 20). These results clearly suggest that the combination of number of oxyethylene units and the shape of the ether cavity leads to this suitable Li^+ selective ionophore.

^{1}H-NMR titration of **34c** and **36a** with Li^+ or Na^+ clearly indicated that the complexes are composed of 1:1 molar ratio of the two components. This experiments also gave an important information on their metal ion binding sites. In Table 4 are summarized chemical shift differences of oxyethylene units between those recorded in the presence or absence of metal salts. The largest shifts were observed for Hb in both experiments with **34c** and **36a**, especially

Fig. 2. Initial conditions of the transport experiments

Source phase	Liquid membrane	Receiving phase
LiOH 0.1 M	Crownophane:	HCl 0.1 M
NaOH 0.1 M	1.5×10^{-4} mol	
KOH 0.1 M	Dodecanoic acid:	
HCl 0.2 M		
(H_2O, 15 ml)	($CHCl_3$, 30 ml)	(H_2O, 15 ml)

Table 3. Transport of alkali metal cations by crownophane–dodecanoic acid cooperative carriers[a]

Crowno-phane	Transport rate ($\times 10^{-6}$ mol/h)[b]			Selectivity	
	Li^+	Na^+	K^+	Li^+/Na^+	Li^+/K^+
35	1.88	0.41	0.46	4.6	4.1
36a	2.01	0.23	0.27	8.6	7.6
34b	0.69	0.37	0.21	1.9	3.2
34c	1.86	0.26	0.41	7.2	4.5
34d	1.32	1.33	0.79	1.0	1.7
37	1.65	0.30	0.62	5.5	2.7
38	2.33	0.66	0.62	3.5	3.8

[a] Reproducibility, $\pm$ 10%.

Table 4. ^{1}H-NMR shift data for alkali metal–crownophane complexes[a]

	Chem. shift difference ($\Delta\delta_{EO} = \delta$(complex) $- \delta$(free))					
Metal ion	Crownophane **36a**			Crownophane **34c**		
	H_a	H_b	H_c	H_a	H_b	H_c
Li$^+$	0.02	0.09	0.06	0.02	0.03	0.02
Na$^+$	0.05	0.06	0.06	0.02	0.02	0.00
K$^+$	0.02	0.02	0.02	0.00	0.01	$-$0.02
Cs$^+$	0.01	0.01	0.01	0.00	0.00	0.00

[a] Recorded on a Varian Gemini-200 FT NMR spectrometer. Experimental conditions: [MBF$_4$]/[**36a** or **34c**] = 5; in CD$_3$CN. Resolution, 0.01 Hz.

when lithium salt was added. Consequently, the metal cations in these complexes are located at the corner of the oxyethylene ring units farthest from the aromatic nuclei. The structures of these complexes are also supported by the framework examination which clearly shows the crownophanes taking the shape of a spoon with the aromatic part as a bail.

The transport rates are considered to be controlled by the extraction and/or release processes. Generally the extraction process is responsible for the rate. For the present crownophane cases, we were able to determine which process controls this transport, carrying out the competitive extraction of alkali metal ions by the crownophane-dodecanoic acid systems. The competitive extraction experiments show the high Li$^+$ selectivity, so that the extraction step is concluded, actually, to control the transport. Note that the Li$^+$ selectivity toward Na$^+$ in the extraction is higher than 10 for each system, as shown in Table 5.

3.4 Behavior of Metacyclophanes During Alkali Metal Cation Transport [79]

Metacrownophane **36a** (Structure 20) showed the most efficient and the highest selectivity toward Li$^+$ among the crownophanes, as shown in Tables 3 and 5. This result was not easily understood. Only this metacrownophane has large conformational mobility between *syn*- and *anti*-conformers. This mobility was thought to be responsible for the high selectivity and efficiency. In order to clarify the effect of conformation on transport, we prepared a dimethoxy derivative of metacrownophane **36b** (s. Table 6), which keeps *anti*-conformation because of the steric hindrance for the conformational change, and examined its transport behavior as well as ^{1}H-NMR titration.

^{1}H-NMR spectra provide significant information for metacrownophane conformational change, based on Lehner's criterion [80] that *anti*-metacyclophanes show negative but large $\Delta\delta$ ($\Delta\delta = \delta H_i - \delta H_e$; s. Table 6 for the proton designation), while *syn*-isomers show a positive and small $\Delta\delta$ value. Although the change in the value of $\Delta\delta$ is virtually insignificant on the addition of KBF$_4$, the addition of LiBF$_4$ or NaBF$_4$ resulted in a decrease in the absolute $\Delta\delta$ values,

Table 5. Competitive extraction of alkali metal ions by crownophane–dodecanoic acid system[a]

Crowno-phane	Decrease of ions in aqueous solution (%)						Selectivity (Li$^+$/M$^+$)			
	Li$^+$	Na$^+$	K$^+$	Rb$^+$	Cs$^+$	Total	Na$^+$	K$^+$	Rb$^+$	Cs$^+$
35	25.2	1.3	4.1	5.5	6.1	41.8	19.4	6.1	4.9	4.1
36a	29.4	1.2	4.3	6.0	7.9	48.8	24.5	6.8	4.9	3.7
34a	19.1	1.6	2.3	3.8	4.6	31.4	11.9	8.3	5.0	4.2
34b	27.9	1.1	4.4	4.6	4.8	42.8	25.4	6.3	6.1	5.8
34c	28.3	1.9	4.6	5.3	5.5	45.6	14.9	6.2	5.3	5.1
34d	27.9	1.2	4.6	4.8	6.5	45.0	23.3	6.1	5.8	4.3

[a] Experimental conditions: composition of cations in 1.0 ml of intial aqueous layer; [LiOH] = [NaOH] = [KOH] = 0.1 mol dm^{-3}, [RbCl] = [CsCl] = 0.1 mol dm^{-3}, [HCl] = 0.2 mol dm^{-3}. Composition of organic layer; 0.05 mmol of crownophane-dodecanoic acid 1:1 mixture in 2.0 ml of chloroform. Reproducibility, ± 10%.

Table 6. ^{1}H-NMR data of metacrownophane **36a** and dimethoxy-metacrownophane **36b** in the presence or absence of M$^+$ (in CD$_3$CN)

36a, 36b

	$\Delta\delta = \delta\text{Hi} - \delta\text{He}$			
	Without M$^+$	Li$^+$	Na$^+$	K$^+$
36a	− 0.40	− 0.32	− 0.26	− 0.39
36b	− 0.18	− 0.50	− 0.71	− 0.28

even though the values were negative. These facts clearly suggest the increased contribution of *syn*-conformation by adding the salts. This phenomenon can be correlated with the salt effect on chemical shift changes of its oxyethylene parts as shown in Table 4. Accordingly, metacrownophane **36a** changes its conformation from the thermodynamically stable *anti*- to *syn*-species. The latter is certainly favorable to the complexation. This can be said to be a kind of induced fit. In contrast, salt addition to dimethoxymetacrownophane **36b** did not affect any chemical shift change of the oxyethylene part. The $\Delta\delta$ value of **36b** changed to a more negative value with the addition of salts. This change is attributed to some kind of solvent polarity effect.

Transport experiments were carried out under the same conditions as mentioned above (s. Fig. 2). As listed in Table 7 metacrownophane **36a**

Table 7. Transport of alkali metal cations by meta-crownophanes with dodecanoic acid

Crowno-phane	Transport rate ($\times 10^{-6}$ mol/h)		
	Li$^+$	Na$^+$	K$^+$
36a	2.01	0.23	0.27
36b	1.16	0.18	0.38

accelerates the Li$^+$ transport about two times more than that of dimethoxy-metacrownophane **36b**. This depression is also explained by the lack of the induced fit for **36b**.

3.5 Synthesis of Double-Looped Crownophanes and Their Ability to Transport Amino Acids through Liquid Membranes [81]

Double-looped crownophanes **39** and **40** were conveniently synthesized by [2 + 2] photocycloaddition as shown in Schemes 2 and 3, respectively. The

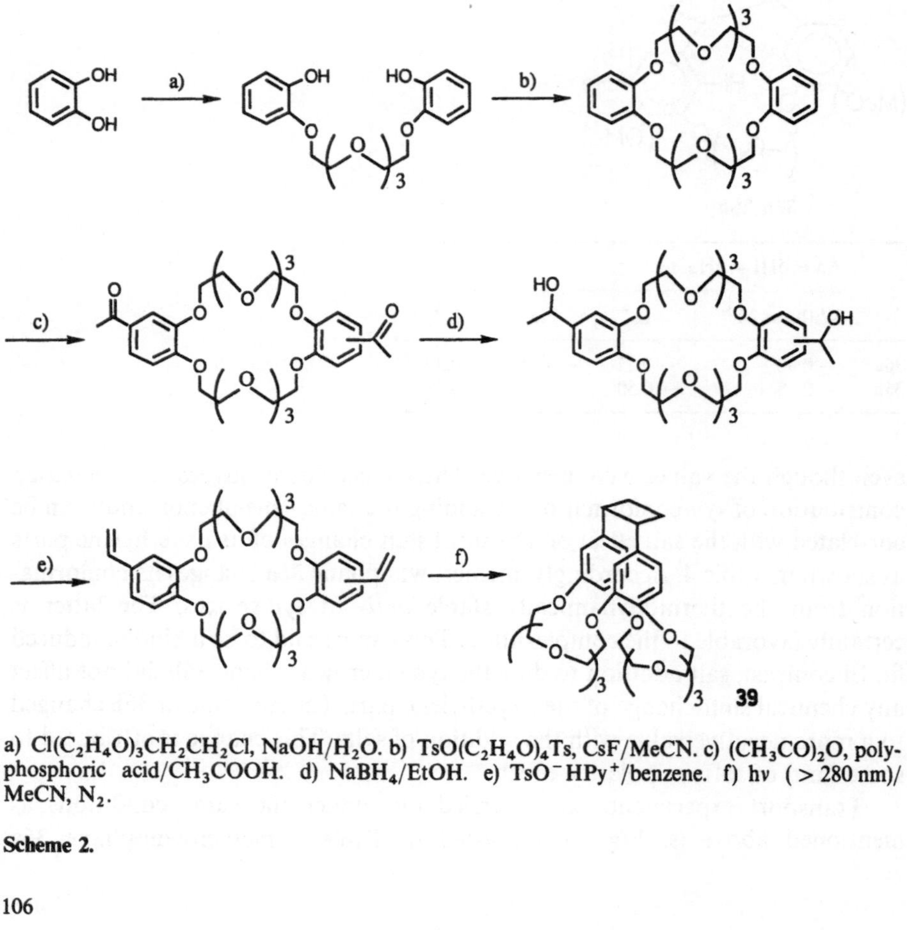

a) Cl(C$_2$H$_4$O)$_3$CH$_2$CH$_2$Cl, NaOH/H$_2$O. b) TsO(C$_2$H$_4$O)$_4$Ts, CsF/MeCN. c) (CH$_3$CO)$_2$O, poly-phosphoric acid/CH$_3$COOH. d) NaBH$_4$/EtOH. e) TsO$^-$HPy$^+$/benzene. f) hv (> 280 nm)/MeCN, N$_2$.

Scheme 2.

Table 8. Competitive passive transport by crownophane-dodecanoic acid cooperative carriers[a]

Crowno-phane	Transport rate ($\times 10^{-6}$ mol/h)				
	Li$^+$	Na$^+$	K$^+$	Rb$^+$	Cs$^+$
34c	3.91	0.09	1.58	0.20	1.22
36a	2.01	0.23	0.27	–	–
39	1.14	0.40	1.20	0.21	1.17
40	5.10	1.33	1.97	2.27	1.55

[a] Experimental conditions: source phase (H$_2$O, 15 ml); [LiOH] = [NaOH] = [KOH] = [RbCl] = [CsCl] = 0.1 M, [HCl] = 0.2 M. Receiving phase (H$_2$O, 15 ml); [HCl] = 0.1 M. Liquid membrane phase (CHCl$_3$, 30 ml); [crownophane] = [dodecanoic acid] = 1.5×10^{-4} mol. Reproducibility, $\pm$ 15%.

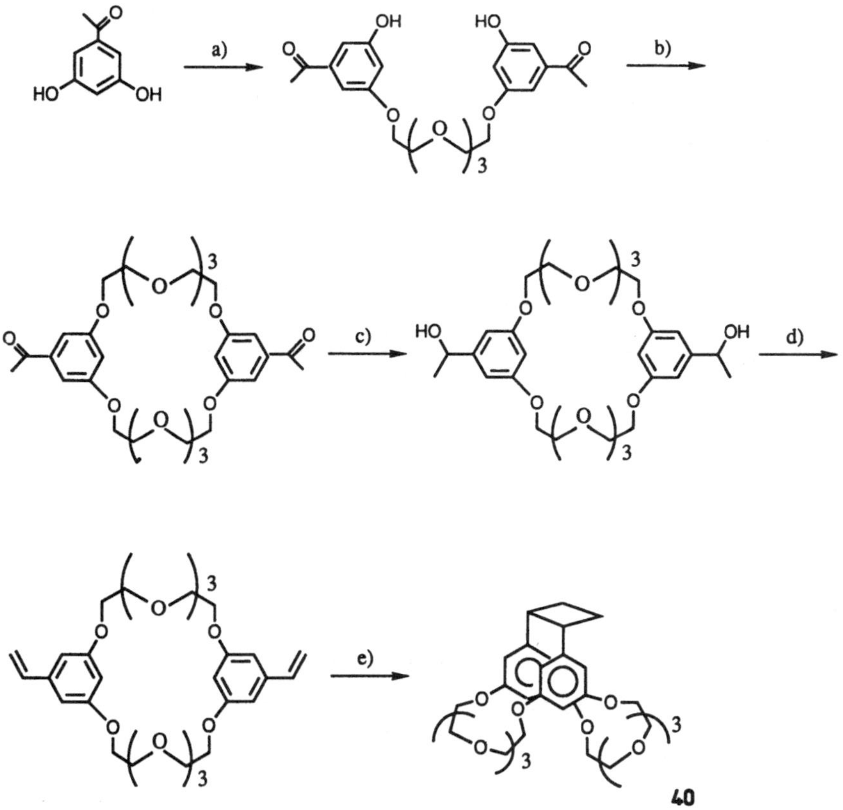

a) Cl(C$_2$H$_4$O)$_3$CH$_2$CH$_2$Cl, NaOH/H$_2$O. b) TsO(C$_2$H$_4$O)$_4$Ts, CsF/MeCN. c) NaBH$_4$/EtOH. d) KHSO$_4$/DMSO. e) hν (> 280 nm)/MeCN, N$_2$

Scheme 3.

double-looped crownophane **40** having the structural element of **36a** facilitated, with dodecanoic acid, the transport of Li^+ selectively in contrast to **39** which did not transport Li^+ effectively.

The Li^+ transport ability of **40** reached almost twice the amount of that of **36a**, suggesting that **40** forms a 1:2 complex in the membrane phase. This result indicates that **40** can transport some diionic species by binding them into its two sites. We were prompted to examine an amino acid transport, taking **40** as a receptor of the two point recognition. The transport experiment was performed by using a cylindrical cell under the conditions shown in Fig. 3.

Double-looped crownophane **40** transported dibasic amino acid derivatives (BzGlu) three-times better than monobasic one such as BzGly. This result can be explained by the increased lipophilicity of the crownophane **40**-dibasic amino acid complexes, due to the simultaneous two-point electrostatic bonding between the two positively-charged crown moieties of **40** and two carboxylate groups. Their high lipophilicity results in the high solubility in the $CHCl_3$ membrane and, at the same time, in a high transport rate for the amino acid anions.

According to the CPK-model examination, the distance (ca. 8.4 Å) of the two crown moieties in the structure of **40** fits the length between two carboxylates in relaxing BzGlu (ca. 8.6 Å) better than that of BzAsp (ca. 7.2 Å). This size complementarity certainly agrees with the transport rate of BzGlu which is

Fig. 3. Initial conditions of the transport for amino acid derivatives and dicarboxylic acids

Source phase	Membrane phase	Receiving phase
Amino acid derivative or dicarboxylic acid 0.1 M	Crownophane:	
LiOH 0.1 M	0.1 mmol	
LiCl 1.0 M		
(H_2O, 10 ml)	($CHCl_3$, 75 ml)	(H_2O, 60 ml)

Table 9. Passive transport of benzoyl amino acid anions by crownophanes[a]

Carriers	Transport rate ($\times 10^{-8}$ mol/h)		
	BzGly (−)	BzAsp (7.2 Å)[b]	BzGlu (8.6 Å)[b]
40	0.74	1.18	2.57
34c[c]	1.35	0.92	0.60
Benzo-15-crown-5[c]	0.64	0.60	0.54

[a] Reproducibility, ± 15%
[b] Inter-carboxylate distance.
[c] The crown unit concentration is half that of double-looped crownophane **40**

Table 10. Passive transport of dicarboxylic acid by double-looped crowno-phane 40[a)]

Dicarboxylic acid	Inter-carboxylate distance (Å)	Transport rate ($\times 10^{-8}$ mol/h)
BzGly	–	0.74
BzAsp	7.2	1.18
BzGlu	8.6	2.57
o-PA	5.0	1.57
m-PA	8.7	1.28
p-PA	10.3	0.40
PhSu	7.2	2.54
PhGl	8.6	4.41

[a)] Reproducibility, $\pm$ 15%.

twice as high as that of BzAsp (s. Table 9). This size complementarity also worked for the transport of aliphatic dibasic acids, such as 2-phenylsuccinic acid (PhSu), and 3-phenylglutaric acid (PhGl), as summarized in Table 10.

Based on the complementarity, double-looped crownophane **40** is expected to transport isophthalic acid (m-PA) most efficiently among other isomers with almost the same hydrophilic-lipophilic balance (HLB) values. Results in Table 10 show that the largest transport efficiency was observed for m-PA and phthalic acid (o-PA), and, as expected, the longest terephthalic acid (p-PA) was not transported so well. Consequently, their transport behavior largely depends on the distance between two crown moieties in the host molecules and that of two carboxylate groups in the guest molecules.

3.6 Synthesis and Ionophoric Properties of Dihydroxycrownophanes and Their Derivatives [82]

The treatment of p-vinylphenol with oligoethylene glycol diglycidyl ether in the presence of alkali metal hydroxide afforded new styrene derivatives **41** possessing two hydroxyl groups on the oligo(oxyethylene) linkages (Scheme 4). These derivatives are converted to the corresponding crownophanes **42** by the photocycloaddition method. Crownophanes **42** having two hydroxyl groups were easily converted into doubly armed crownophanes.

Silver is the most important metal in the photographic industries and its complete recovery from the photographic waste stream is very desirable from the points of view of saving resources and preserving the environment [83]. Selective transport or extraction of Pb^{2+} ions is also of interest in relation to the environmental and human toxicity [84]. We were thus stimulated to synthesize some crownophanes possessing alkylthio-, picolyl- or carboxyl-groups in the side chains, expecting them to exhibit such silver and lead extraction abilities.

All the compounds obtained (Scheme 4) were the subject of solvent extraction experiments of metal nitrates from aqueous solution into CH_2Cl_2 solution. Results are summarized in Table 11.

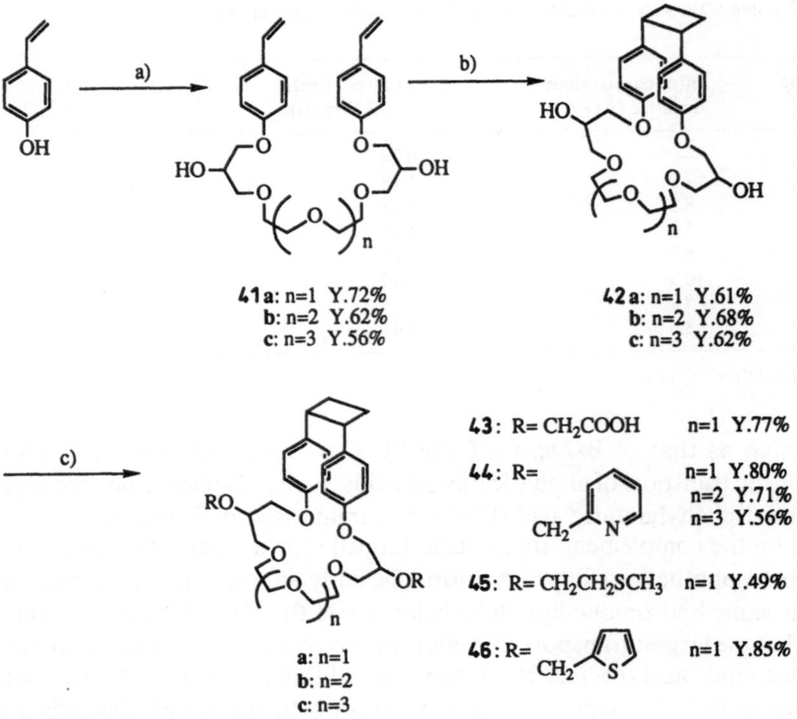

41a: n=1 Y.72%
  b: n=2 Y.62%
  c: n=3 Y.56%

42a: n=1 Y.61%
  b: n=2 Y.68%
  c: n=3 Y.62%

43: R= CH$_2$COOH   n=1 Y.77%

44: R= ... CH$_2$-N   n=1 Y.80%
                            n=2 Y.71%
                            n=3 Y.56%

45: R= CH$_2$CH$_2$SCH$_3$ n=1 Y.49%

46: R= CH$_2$- S   n=1 Y.85%

a: n=1
b: n=2
c: n=3

Scheme 4.

Table 11. Extraction of metal nitrates with crownophanes bearing two ligands on the polyether moieties[a]

Ligands	Percent extraction							
	Ag$^+$	Cu^{2+}	Pb^{2+}	Ni^{2+}	Zn^{2+}	Co^{2+}	Mn^{2+}	Fe^{3+}
42a	5(4.2)	3(4.2)	4(4.8)	11(6.6)	11(6.6)	0(5.8)	0(5.8)	7(2.0)
43a	43(4.0)	54(4.1)	82(4.7)	18(6.7)	35(5.4)	6(6.2)	18(6.2)	12(2.0)
44a	87(4.3)	0(4.2)	2(4.8)	3(6.8)	2(5.9)	0(6.7)	0(6.7)	14(2.0)
44b	84(4.2)	0(4.2)	4(4.7)	0(6.7)	0(5.2)	0(5.9)	0(5.8)	0(2.0)
44c	77(4.4)	0(4.2)	1(4.7)	3(6.9)	0(5.5)	0(6.2)	0(6.2)	0(1.8)
45a	0(4.3)	0(4.2)	2(4.8)	3(6.8)	1(5.9)	0(6.6)	0(6.6)	7(2.0)
46a	3(4.3)	0(4.3)	0(4.5)	0(6.9)	3(6.0)	0(5.2)	0(5.2)	0(2.0)

[a] Extraction condition: Aq. phase, [metal nitrate] $= 1 \times 10^{-1}$ M, 5 ml; org. phase, CH$_2$Cl$_2$, [ligand] $= 1 \times 10^{-4}$ M, 5 ml. Values in parentheses are equilibrium pH of the aqueous phase.

Although dihydroxycrownophane **42a** did not extract any cations significantly, crownophane dicarboxylic acid **43a** efficiently extracted almost all heavy metal cations examined in this work. The Pb^{2+} ion was extracted most efficiently among the divalent cations, and the extraction percentage was higher than that of Ag$^+$ in spite of the similar ionic diameter (Pb^{2+}, 2.38 Å; Ag$^+$,

2.32 Å). The binding behavior of **43a** to Ag$^+$ and Pb^{2+} was studied by a continuous variation method in the extraction experiments (s. Fig. 4a and b). Percent extractions of both cases reached to maximum at 0.5 molar fraction. This fact clearly indicates that both Ag$^+$ and Pb^{2+} form 1:1 complexes with **43a**, and also suggests that the Pb^{2+} and Ag$^+$ ions fit into the cavity of **43a**. The neutralization of the charges between the host and metal cation must occur more favorably with Pb^{2+} than with Ag$^+$, because 1:1 complexes must exist in the organic phase. Accordingly, it is considered that the **43a**-Pb^{2+} complex

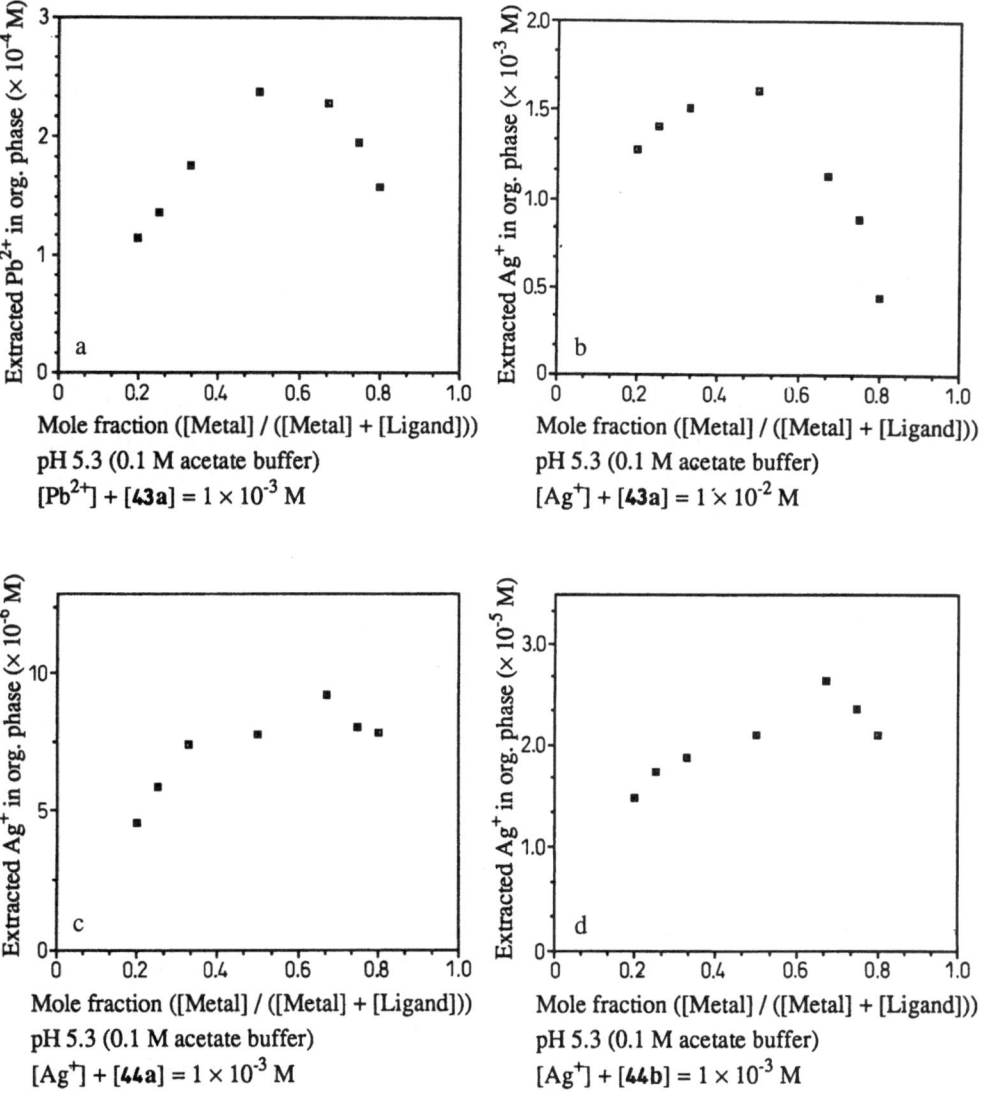

Fig. 4. Job plot of the extraction of cations by crownophanes

becomes more lipophilic than the **43a**-Ag$^+$ one, so that the former is more easily extracted into the CH$_2$Cl$_2$ phase than the latter. In fact, the Pb^{2+} ion was extracted with **43a** more efficiently than Ag$^+$ ion.

It is clearly revealed that **43a** is a useful ligand for recovering heavy and transition metal cations, especially Ag$^+$ and Pb^{2+}, from neutral water or sea water which contains a plenty of alkali and alkaline metal cations, because **43a** did not extract any alkali and alkaline earth metal cation even at pH 7.

Cyclic and acyclic sulfide moieties in the side chains are believed to act as the covers on the crownophane cavities rather than cation-ligating groups since crownophane derivatives **45a** and **46a** could not extract any of the cations examined [85].

Crownophanes **44** possessing two pyridine moieties in the side chains showed excellent selectivities and high efficiencies as regards Ag$^+$ ions. These results can be explained from the specific affinity [83, 86] between nitrogen atoms of pyridine rings and Ag$^+$ ions, and also from the high lipophilicity of the complexes composed of single charged Ag$^+$ ions and the host molecules, compared with other complexes formed from crownophanes and plural-charged cations which require two or three hydrophilic nitrate anions. It is suggested that the picolyl groups on the side chains play a significant role in binding Ag$^+$ ions by making intramolecular cooperation between the picolyl groups and the polyether moiety. This trend can be seen from the results of the similar extractability of **44a**, **44b**, and **44c** toward Ag$^+$ ions.

The binding behavior of **44a** and **44b** to Ag$^+$ ions was determined by the continuous variation method (Fig. 4c and d). The maxima are seen at molar fractions larger than 0.5. Moreover, these figure shapes are much different from those of Fig. 4b. Although several kinds of complexes having various host/guest ratios are actually formed, the ratios have not been determined yet.

3.7 Synthesis and Binding Properties of New Doubly Armed Crownophanes [87]

A variety of crown compounds possessing cation-ligating side chain(s) have been synthesized while searching for more specific and/or more efficient complexing ability than with the parent compounds. These kinds of functional crown compound were named "lariat ethers" [88–95] and "BiBLEs" [96, 97] by Gokel and co-workers. Okahara and co-workers [78, 98–104] and Inokuma [104–110] have also reported the relationships between the structures and alkali metal-complexing abilities for some lariat ethers and BiBLEs. Tsukube and co-workers have reported specific cation-binding properties of a series of "double-armed crown ethers" [85, 112–118].

In this section the synthesis (Scheme 5) and cation-binding properties of doubly pyridine-armed crownophane **48** are dealt with. Cation binding properties of **48** were examined by the liquid-liquid extraction method. Results are summarized in Table 12 together with some reference compounds.

Although parent compound **47** (s. Scheme 5) and linear reference compounds **49** and **50** (Table 12) which have two picolyl groups showed low extractabilities toward Ag^+, doubly armed crownophane **48** efficiently extracted Ag^+, and with excellent selectivity. Ouchi and co-workers reported that a pair of polyether side chains attached at the same carbon atom of a crown ether can interact

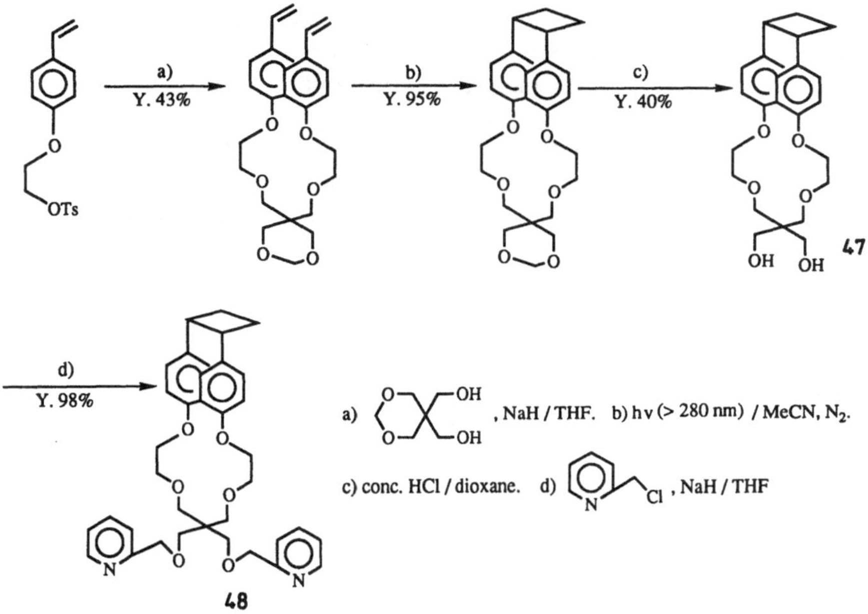

Scheme 5.

Table 12. Extraction of metal nitrates with ligands

Ligand	Extractability (%)[a]						
	Ag^+	Cu^{2+}	Pb^{2+}	Ni^{2+}	Zn^{2+}	Co^{2+}	Mn^{2+}
47	10(3.4)	0(4.7)	0(3.9)	0(6.9)	0(6.0)	0(7.1)	0(6.0)
48	78(3.0)	4(4.4)	0(4.9)	0(6.9)	0(5.7)	0(6.9)	0(5.7)
49	3(3.4)	0(4.5)	0(4.9)	0(6.9)	0(6.1)	0(7.1)	0(6.7)
50	3(3.4)	0(4.5)	0(4.9)	0(6.9)	0(6.2)	0(7.2)	0(6.8)

[a] Extraction conditions: Aq. phase, [metal nitrate] = 1×10^{-1} mol dm^{-3}, 5 ml; Org. phase, CH_2Cl_2, [ligand] = 1×10^{-4} mol dm^{-3}, 5 ml.
Values in parentheses are equilibrium pH of aqueous phase.

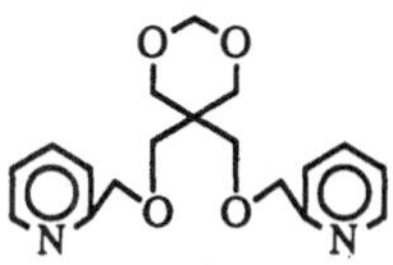

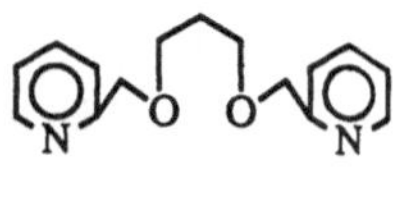

with various kinds of cations, but both side chains cannot work simultaneously with the crown ether ring, according to the consideration of bond-angles [119]. Therefore, our results mentioned above are considered to be caused by the synergism of crownophane moieties and one side chain and this increases the complexation of **48** and Ag$^+$. Binding behavior of Ag$^+$ to **48** was estimated by the continuous variation method in the extraction experiments (Figure 5). Percent extractions reached a maximum at a 0.5 mole fraction for this cation. The fact clearly indicates that Ag$^+$ forms a 1:1 complex with **48**.

The pH limitation for the extraction was examined from the practical point of view. As shown in Fig. 6, the high Ag$^+$ extractability of **48** was retained even at low pH values (pH $\approx$ 1) of a contacting aqueous phase (equilibrium pH). The reference compounds **49** and **50**, however, always exhibited very low extractability. The results clearly suggest that the new crownophane **48** is a useful ionophore for recovery of Ag$^+$ ions not only from natural or sea water but also from waste streams [120] generally with low pH values.

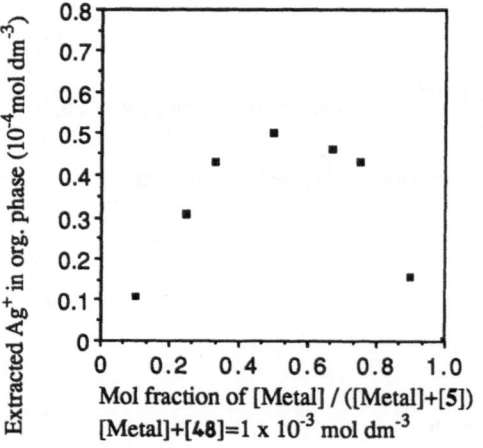

Fig. 5. Continuous variation plot of the extraction of Ag$^+$ with host **48**

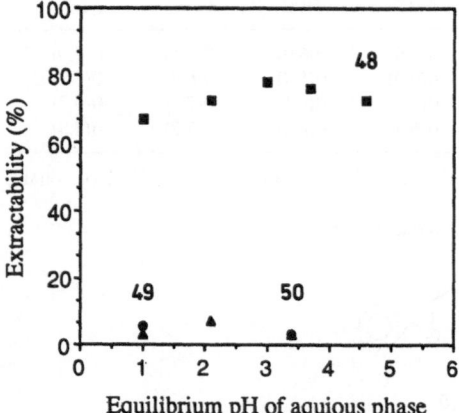

Fig. 6. pH Effect on the extraction of Ag$^+$

4 Conclusion

Cyclophane chemistry offers intriguing and important structural building blocks to design desirable compounds in host-guest chemistry. Many attempts, which have been mentioned in this review and have recently been published in the chemical literature [121–124] were made to hybridize some ion ligating functional rings, moieties, or groups with the characteristic cyclophane skeletons. In this trend, our synthetic method has some advantages. Firstly, it is useful for designing molecules as small as possible in order to enable easy diffusion through a membrane. Secondly, the molecules can be made amphiphilic, because of the proper proportion of hydrophilic and lipophilic parts which can be easily attained. Thirdly, the solubility of intermediates toward designed molecules can be retained in its synthetic sequence, because the cyclization process can be carried out at the very end. More properly speaking, cyclophanes normally have low solubility in conventional solvents. Fourthly, many preparation methods now exist for the introduction of vinyl groups into aromatic rings, such as the old fashioned dehydration of phenethyl alcohol moieties [125] or the recently developed Stille reaction [126]. We are hopeful that all these methods will be used to make receptors and ionophores, that show high selectivity and efficiency to some important natural products as well as valuable or toxic minerals.

5 References

1. Nishimura J, Doi H, Ueda E, Ohbayashi A, Oku A (1987) J Am Chem Soc 109: 5293
2. Nishimura J, Ohbayashi A, Wada Y, Oku A, Ito S, Tsuchida A, Yamamoto M, Nishijima Y (1988) Tetrahedron Lett 29: 5375
3. Nishimura J, Ohbayashi A, Doi H, Nishimura K, Oku A (1988) Chem Ber 121: 2019
4. Nishimura J, Ohbayashi A, Ueda E, Oku A (1988) Chem Ber 121: 2025
5. Nishimura J, Takeuchi M, Takahashi H, Ueda E, Matsuda Y, Oku A (1989) Bull Chem Soc Jpn 62: 3161
6. Nishimura J, Horikoshi Y, Takahashi H, Machino S, Oku A (1989) Tetrahedron Lett 30: 5439
7. Okada Y, Sugiyama K, Wada Y, a Nishimura J (1990) Tetrahedron Lett 31: 107
8. Nishimura J, Takeuchi M, Takahashi H, Sato M (1990) Tetrahedron Lett 31: 2911
9. Nishimura J, Horikoshi J, Wada Y, Takahashi H, Sato M (1991) J Am Chem Soc 113: 3485
10. Nishimura J, Wada Y, Sano Y (1992) Bull Chem Soc Jpn 65: 618
11. Nishimura J, Horikoshi Y (1992) Bull Chem Soc Jpn 65: 941–942
12. Takeuchi M, Nishimura J (1992) Tetrahedron Lett 33: 5563
13. Wada Y, Ishimura T, Nishimura J (1992) Chem Ber 125: 2155
14. Nishimura J (1992) In: Osawa E (ed) Carbocyclic cage compounds. Verlag Chem, Stuttgart, p 156
15. Nishimura J, Takeuchi M, Koike M, Yamashita O, Okada J, Takaishi M (1993) Bull Chem Soc Jpn 66: 598
16. Takeuchi M, Tuihiji, T, Nishimura J (1993) J Org Chem 53: 7388
17. Takeuchi M, Inokuma S, Nishimura J (1993) Yuki Gousei Kyoukaishi 52: 652
18. Inokuma S, Nishimura J (1993) Yuki Gousei Kyoukaishi 52: 673
19. Inokuma S, Yabusa K, Kuwamura T: Chem Lett 1984: 607
20. Inokuma S, Hayase T, Yabusa K, Ashizawa T, Kuwamura T: Nippon Kagaku Kaishi 1987: 1059
21. Inokuma S, Hayase T, Azechi S, Kuwamura T: Nippon Kagaku Kaishi 1988: 662

22. Kawashima N, Kawashima T, Otsubo T, Misumi S (1978) Nippon Kagaku Kaishi 5025
23. Heleson RC, Timko JM, Cram DJ (1974) J Am Chem Soc 96: 7380
24. Heleson RC, Tarnowski TL, Timko JM, Cram DJ (1977) J Am Chem Soc 99: 6411
25. Allwood BL, Spencer N, Shahriari-Zavarech H, Stoddart JF, Williams DJ: J Chem Soc, Chem Commun 1987: 1061
26. Allwood BL, Spencer N, Shahriari-Zavarech H, Stoddart JF, Williams DJ: J Chem Soc, Chem Commun 1987: 1064
27. Ashton PR, Slawin AMZ, Spencer N, Stoddart JF, Williams DJ: J Chem Soc, Chem Commun 1987: 1066
28. Allwood BL, Kohnke FH, Stoddart JF, Williams DJ (1985) Angew Chem Int Ed Engl 24: 581
29. Allwood BL, Shahriari-Zavarech H, Stoddart JF, Williams DJ (1987) Angew Chem Int Ed Engl 26: 1058
30. Frensch K, Vögtle F (1979) J Org Chem 44: 884
31. Jarvi ET, Whitlock Jr. HW (1980) J Am Chem Soc 102: 657
32. Whitlock BJ, Jarvi ET, Whitlock HW (1981) J Org Chem 46: 1832
33. Adams SP, Whitlock Jr. HW (1981) J Org Chem 46: 3474
34. Adams SP, Whitlock HW (1982) J Am Chem Soc 104: 1602
35. Gabard J, Collet A: J Chem Soc, Chem Commun 1981: 1137
36. Canceill J, Lacombe L, Collet A: J Chem Soc, Chem Commun 1987: 219
37. Canceill J, Lacombe L, Collet A (1985) J Am Chem Soc 107: 6993
38. Canceill J, Lacombe L, Collet A (1986) J Am Chem Soc 108: 4230
39. Collet A (1987) Tetrahedron 43: 5725
40. Canceill J, Cesario M, Collet A, Guilhem J, Riche C, Pascard C: J Chem Soc, Chem Commun 1985: 361
41. Canceill J, Cesario M, Collet A, Guilhem J, Riche C, Pascard C: J Chem Soc, Chem Commun 1986: 339
42. Brown GR, Chana SS, Stoddart JF, Slawin AMZ, Williams DJ: J Chem Soc, Perkin Trans 1 1989: 211
43. Bower GR, Slawin AM, Williams DJ, Brown GR, Chana SS, Stoddart JF: J Chem Soc, Perkin Trans 1 1989: 212
44. Alfieri C, Dradi E, Pochini A, Ungaro R, Andreetti GD: J Chem Soc, Chem Commun 1983: 1075
45. Reinhoudt DN, Dijkstra PJ, in't Veld PJA, Bugge KE, Harkema S, Ungaro R, Ghidini E (1987) J Am Chem Soc 109: 4761
46. van Loon J-D, Arduini A, Verboon W, Ungaro R, van Hummel GJ, Harkema S, Reinhoudt DN (1989) Tetrahedron Lett 20: 2681
47. Dijkstra PJ, Brunink JAJ, Bugge K-E, Reinhoudt DN, Harkema S, Ungaro R, Ugozzoli F, Ghidini E (1989) J Am Chem Soc 111: 7567
48. Ghidini E, Ugozzoli F, Ungaro R, Harkema S, El-Fadl AA, Reinhoudt DN (1990) J Am Chem Soc 112: 6979
49. Arduini A, Casnati A, Dodi L, Pochini A, Ungaro R: J Chem Soc, Chem Commun 1990: 1597
50. van-Loon J-D, Groenen LC, Wijmenga SS, Verboom W, Reinhoudt DN (1991) J Am Chem Soc 113: 2378
51. Cram DJ, Kaneda T, Helgeson RC, Lein GM (1979) J Am Chem Soc 101: 6754
52. Cram DJ, Lein GM (1985) J Am Chem Soc 107: 3657
53. Trueblood KN, Knobler CB, Maverick E, Helgeson RC, Brown SB, Cram DJ (1981) J Am Chem Soc 103: 5594
54. Cram DJ, Lein GM, Kaneda T, Helgeson RC, Knobler CB, Naverick E, Trueblood KN (1981) J Am Chem Soc 103: 6228
55. Cram DJ (1986) Angew Chem Int Ed Engl 25: 1039
56. Artz SP, Cram DJ (1984) J Am Chem Soc 106: 2160
57. Cram DJ, Dicker IB, Lein GM, Knobler CB, Trueblood KN (1982) J Am Chem Soc 104: 6827
58. Katz HE, Cram DJ (1984) J Am Chem Soc 106: 4977
59. Dijkstra PJ, van Steen BJ, Hams BHM, den Hertog Jr. HJ, Reinhoudt DN (1986) Tetrahedron Lett 27: 3183
60. Dijkstra PJ, den Hertog HJ, van Steen BJ, Zijlstra S, Skowronska-Ptasinska M, Reinhoudt DN, van Eerden J, Harkema S (1987) J Org Chem 52: 2433
61. Dijkstra PJ, Skowronska-Ptasinska MHJ, Reinhoudt DN, den Hertog HJ Jr., van Eerden J, Harkema S, Zeeuw D (1987) J Org Chem 52: 4913

62. Dijkstra PJ, den Hertog HJ Jr., van Eerden J, Harkema S, Reinhoudt DN (1988) J Org Chem 53: 374
63. Dijkstra PJ, Brunink JAJ, Bugge K-E, Reinhoudt DN, Harkema S, Ungaro R, Ugozzoli F, Ghidini E (1989) J Am Chem Soc 111: 7567
64. Ashton PR, Chrystal EJT, Mathias JP, Parry KP, Slawin AMZ, Spencer N, Stoddart JF, Williams DJ (1987) Tetrahedron Lett 28: 6367
65. Ortholand J-Y, Slawin AMZ, Spencer N, Stoddart JF, Williams DJ (1989) Angew Chem Int Ed Engl 28: 1394
66. Bouas-Laurent H, Castellan A, Daney M, Desvergne J-P, Guinand G, Marsau P, Riffaud M-H (1986) J Am Chem Soc 108: 315
67. Guinand G, Marsau P, Bouas-Laurent H, Castellan A, Desvergne J-P, Lamotte M (1987) Acta Crystallogr, Sect C 43: 857
68. Fages F, Desvergne J-P, Bouas-Laurent H, Lehn, J-M, Konopelski JP, Marsau P, Barrans Y: J Chem Soc, Chem Commun 1990: 655
69. Shanzer A, Libman J, Frolow F (1983) Acc Chem Res 16: 60
70. Hinschberger J, Desvergne J-P, Bouas-Laurent H, Marsau P: J Chem Soc, Perkin Trans 2, 1990: 993
71. Kuo P-L, Miki M, Okahara M: J Chem Soc, Chem Commun 1978: 504
72. Inokuma S, Yamamoto T, Nishimura J (1990) Tetrahedron Lett 31: 97
73. Kuzuya M, Tanaka M, Hosoda M, Noguchi A, Okuda T: Nippon Kagaku Kaishi 1984: 22
74. Hiratani K, Taguchi K, Sugihara H, Iio K (1984) Bull Chem Soc Jpn 57: 1976
75. Tosteson DC (1981) Sci Am 244: 164
76. Charewicz WA, Heo GS, Bartsch RA (1982) Anal Chem 54: 2094
77. Sakamoto H, Kimura K, Shono T (1987) Anal Chem 59: 1513
78. Nakatsuji Y, Wakita R, Harada Y, Okahara M (1989) J Org Chem 54: 2988
79. Inokuma S, Sakai S, Katoh R, Yasuda T, Nishimura J, Nippon Kagaku Kaishi, 1993: 1148
80. Krois D, Lehner H (1982) Tetrahedron 38: 3319
81. Inokuma S, Sakai S, Yamamoto T, Nishimura J: J Membr Sci (in press)
82. Inokuma S, Sakai S, Katoh R, Yasuda T, Nishimura J (1994) Bull Chem Soc Jpn 67: 1462
83. Izatt RM, Lind DH, Bruening RL, Hunszthy P, McDaniel CW, Bradshaw JC, Christensen JJ (1988) Anal Chem 60: 1694
84. Lamb JD, Izatt RM, Robertson PA, Christensen JJ (1980) J Am Chem Soc 102: 2452
85. Tsukube H, Yamashita K, Iwachido K, Zenki M (1989) Tetrahedron Lett 30: 3983
86. Kumar S, Singh R, Harjit S: J Chem Soc, Perkin Trans 1, 1992: 3049
87. Inokuma S, Yasuda T, Araki S, Sakai S, Nishimura J: Chem Lett 1994: 201
88. Gokel GW, Dishong DM, Diamond CJ: J Chem Soc, Chem Commun 1980: 1053
89. Dishong DM, Diamond CJ, Gokel GW (1981) Tetrahedron Lett 22: 1663
90. Schultz RA, Dishong DM, Gokel GW (1981) Tetrahedron Lett 22: 2623
91. Schultz RA, Dishong DM, Gokel GW (1982) J Am Chem Soc 104: 625
92. Schultz RA, Schlegel E, Dishong DM, Gokel GW: J Chem Soc, Chem Commun 1982: 242
93. Dishong DM, Diamond CJ, Cinoman MI, Gokel GW (1983) J Am Chem Soc 105: 586
94. Schultz RA, White BD, Dishong DM, Arnold KA, Gokel GW (1985) J Am Chem Soc 107: 6659
95. Hernandez CJ, Trafton JE, Gokel GW (1991) Tetrahedron Lett 32: 6269
96. Gatto VJ, Gokel GEW (1984) J Am Chem Soc 106: 8240
97. Zinic M, Frkanec L, Skaric V, Trafton J, Gokel GW: J Chem Soc, Chem Commun 1990: 1726
98. Masuyama A, Nakatsuji Y, Ikeda I, Okahara M (1981) Tetrahedron Lett 22: 4665
99. Nakatsuji Y, Nakamura T, Okahara M: Chem Lett 1982: 1207
100. Nakatsuji Y, Nakamura T, Okahara M, Dishong DM, Gokel GW (1982) Tetrahedron Lett 23: 1351
101. Nakatsuji Y, Nakamura T, Okahara M, Dishong DM, Gokel GW (1983) J Org Chem 48: 1237
102. Masuyama A, Kuo P-L, Ikeda I, Okahara M: Nippon Kagaku Kaishi 1983: 249
103. Ikeda I, Emura H, Okahara M (1984) Bull Chem Soc Jpn 57: 1612
104. Nakatsuji Y, Nakamura T, Yonetani M, Yuya H, Okahara M (1988) J Am Chem Soc 110: 531
105. Inokuma S, Kohno T, Inoue K, Yabusa K, Kuwamura T: Nippon Kagaku Kaishi 1985: 1585
106. Inokuma S, Kuwamura T (1987) J Jpn Oil Chem Soc 36: 571
107. Inokuma S, Irisawa Y, Kuwamura T (1988) Yukagaku 37: 33
108. Inokuma S, Itoh D, Kuwamura T (1988) Yukagaku 37: 441
109. Inokuma S, Itoh D, Kobori W, Kuwamura T (1988) Yukagaku 37: 668
110. Inokuma S, Matsunaga T, Negishi, Kuwamura T (1989) Yukagaku 38: 705

111. Inokuma S, Okada H, Kuwamura T (1989) Yukagaku 38: 705
112. Tsukube H: J Chem Soc, Chem Commun 1984: 315
113. Tsukube H, Iwachido T, Hayama N: J Chem Soc, Perkin 1 1986: 1033
114. Tsukube H, Yamashita K, Iwachido T, Zenki M (1988) Tetrahedron Lett 29: 569
115. Tsukube H, Adachi H, Morosawa S: J Chem Soc, Perkin Trans 1989: 1537
116. Matsumoto K, Minatogawa H, Munakata M, Toda M, Tsukube H (1990) Tetrahedron Lett 31: 3923
117. Tsukube H, Yamashita K, Iwachido T, Zenki M (1991) J Org Chem 56: 268
118. Tsukube H, Minatogawa M, Munakata M, Toda M, Matsumoto K (1992) J Org Chem 57: 542
119. Ouchi M, Inoue Y, Wada K, Iketani S, Hakushi T, Weber E (1987) J Org Chem 52: 2420
120. McDowell WJ, Case GN, McDonough JA, Bartsch RA (1992) Anal Chem 64: 3013
121. Gunter M, Johnson MR (1990) Tetrahedron Lett 33: 4801
122. Nijenhuis WF, van Doorn AR, Reichwein AM, de Jong F, Reinhoudt DN (1991) J Am Chem Soc 113: 3607
123. Kraft P, Cacciapaglia R, Böhmer V, El-Fadl AA, Harkema S, Mandolini L, Reinhoudt DN, Verboom W, Vogt W (1992) J Org Chem 57: 826
124. Gunter M, Johnson MR: J Chem Soc, Chem Commun 1992: 1163
125. Nishimura J, Ishida Y, Hashimoto K, Shimizu Y, Oku A, Yamashita S (1981) Polymer J 13: 635
126. Echavarren AM, Stille JK (1978) J Am Chem Soc 109: 5478

Cuppedo- and Cappedophanes

T.K. Vinod[†] and Harold Hart

Department of Chemistry, Michigan State University, East Lansing, MI 48824, USA

Table of Contents

[†] Present address: Department of Chemistry, University of Oregon, Eugene, OR 97402, USA.

Topics in Current Chemistry, Vol. 172
© Springer-Verlag Berlin Heidelberg 1994

The cuppedo- and cappedophanes described here are special cyclophanes whose structures are mainly based on a *m*-terphenyl framework in which substituents cause the outer two rings to be roughly perpendicular to the central ring. With cuppedophanes, two tethers link the outer rings, one in front of and one behind the central ring, to form a cup-shaped or concave molecule. Often some organic functionality is located within the cup, attached to $C_{2'}$ of the *m*-terphenyl moiety. With cappedophanes, an enclosed cavity is formed by attaching the cap via four tethers, two to each of the outer rings. Although there is a strong tendency to form self-filled rather than vaulted cappedophanes, the latter can be obtained provided that the tethers are somewhat rigid and the 5'-position of the *m*-terphenyl moiety carries a bulky substituent. This methodology opens the way for the synthesis of cuppedo- and cappedophanes containing functionality within a well-defined cavity. Synthetic routes to these phanes are discussed in some detail, as are their NMR spectra, conformations, and selective reactions.

1 Introduction

During the last decade various classes of molecules such as cyclodextrins [1], crown ethers [2], calixarenes [3] and cyclophanes [4] have been extensively investigated as abiotic models of enzymes. Among compounds studied from this perspective, investigations of cyclophanes with well defined cavities has been extremely fruitful. Large hydrophobic cavities, heteroatom binding sites on bridges and the possibility of juxtaposing catalytic sites at or near the cavity make cyclophanes a well-tunable model to study "substrate-receptor" interactions in an abiotic enzyme model [5]. Easy synthetic routes can modify the size and shape of cavities of already existing cyclophanes, and the design and synthesis of new classes of cyclophanes significantly aided research along these lines. Our work on cuppedophanes and cappedophanes was initiated with a view of designing two new types of *m*-terphenyl based cyclophanes with cavities (microenvironment) within their molecular framework. A convenient one pot route to various *m*-terphenyls [6] developed in our laboratories during the mid 1980's led us into the cyclophane area described in this review.

2 Architectural Features

2.1 General Structures and Trivial Names

The two new classes of cyclophanes, namely cuppedo- and cappedophanes, are represented by the general formulas **1** and **2** respectively. These structures are based on the *m*-terphenyl template **3**, where the outer two rings are orthogonal to the central ring. The perpendicularity of the ring planes and the slight tilt of the two outer rings towards each other makes **3** a shallow concave base for structures of type **1** and **2**. Tethers between the 2,2″- and 6,6″-positions as in **1** create a molecular bowl or cup [7], the depth of which will depend on the nature of the tethers. Our synthetic route to *m*-terphenyls (vide-infra) allows one to readily incorporate suitable functionality (E) at the $C_{2'}$ position, from hereon referred to as the 'internal' position. In **2**, four tethers connect a capping unit to positions 2,2″, 6,6″-. Depending on the length and rigidity of the tethers a functional group E may be accommodated between the cap and the molecular base. The position of attachment of the tether sites (X) on the molecular template **3** can be displaced one carbon to the 3,3″- and 5,5″-positions to create a cuppedophane of general structure **4** with a slightly larger cavity. A mono-cyclic structure obtained by tethering the 4,4″- positions leads to a new variance and adds the possibility of positioning two suitable (ligating or H-bonding) functionalities facing each other, as in E and E′ shown in structure **5**. The cupped (**1** and **4**) and the capped (**2**) nature of these compounds and their cyclophane

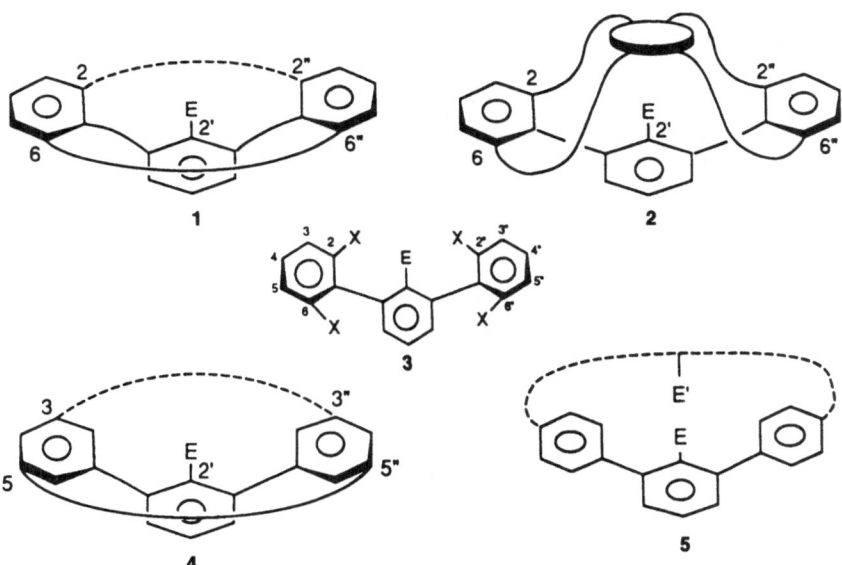

Structures 1

genus prompted us to coin the terms cuppedophanes and cappedophanes to describe these molecules.

The central benzene ring of cuppedophane **1** can be replaced by various other aromatic or heteroaromatic rings (for example, 9,10-phenanthroline) as in general representation **6**. Cuppedophanes of this type possess basic sites in their concave cup. Various members of this class have recently been synthesized and studied by Lüning and co-workers as a part of their effort to design concave reagents [8].

6

Structure 2

2.2 Nature of Tethers and Modes and Linkage

A variety of tethers such as hydrocarbon chains, polyoxamethylene units, and arene units have been employed to construct cuppedophanes and cappedophanes with the general structures noted above. In cuppedophanes, the depth of the cup (that is, the volume of the enclosed cavity) being a critical feature, it is not surprising the arene units, being rigid, were frequently employed as a part of the tether. Several modes of attachment of the tether to the *m*-terphenyl base, to generate cuppedophanes of type **1**, have been worked out. The nature of linkages between the tether and the *m*-terphenyl base depends on the choice of substituent X on **3**. The tetramethyl compound **7** (X = CH₃) allows one to suitably functionalize the four methyl groups to the corresponding tetrakis-bromomethyl compound **8** [9], tetrakis-hydroxymethyl compound **9** [10], tetrakis-mercaptomethyl compound **10** [9] and to tetrakis-tosylamide **11** [11]. Section 3.2.2 describes in detail these various functionalizations. Derivatives **9**, **10**, and **11** can be employed as the *m*-terphenyl base synthon for the construction of

7 (X = CH₃)
8 (X = CH₂Br)
9 (X = CH₂OH)
10 (X = CH₂SH)
11 (X = CH₂NHTs)
12 (X = CH₂NH₂)
13 (X = OH)

Structures 3

oxa-, thia- and aza-cyclophanes respectively. The bromo derivative **8**, apart from frequently serving as precursor to **9, 10,** and **11**, also acts as a *m*-terphenyl synthon in coupling reactions with tether components carrying nucleophilic functionality. The tetraphenol **13** [10] has also been used as a synthon for oxaphanes. *m*-Terphenyl derivatives similar to **8–13** but with all the functional appendages moved one carbon over to the 3,3''- and 5,5''-position are base synthons for cuppedophanes of general structure **4**. Similarly, 4,4''-disubstituted *m*-terphenyl units are useful synthons for macrocycles of general structure **5**.

3 Synthesis of *m*-Terphenyl Templates

3.1 Synthesis of *m*-Terphenyls

A convenient one pot route to various biphenyls [12], *p*-terphenyls [13] as well as *m*-terphenyls [6], starting from suitably substituted polyhalobenzenes, was developed in our laboratories in the mid 1980s. The protocol involves a tandem aryne generation – nucleophilic capture sequence and is described in detail here only in conjunction with the synthesis of *m*-terphenyls. 2,6-Dichloroiodoben-zene **14** is frequently employed as the polyhalobenzene required for the *m*-terphenyl synthesis. It is readily prepared from commercially available 2,6-dichloroaniline via a diazotization sequence. Treatment of **14** with three or more equivalents of arylmagnesium bromide followed by aqueous work-up gives good to excellent yields of *m*-terphenyls. Mechanistic details of this one pot procedure are shown in Scheme 1 and are as follows. The first equivalent of arylmagnesium bromide effects metal–halogen exchange at the iodine site of **14** to give dichlorophenylmagnesium bromide **15**. Elimination of elements of

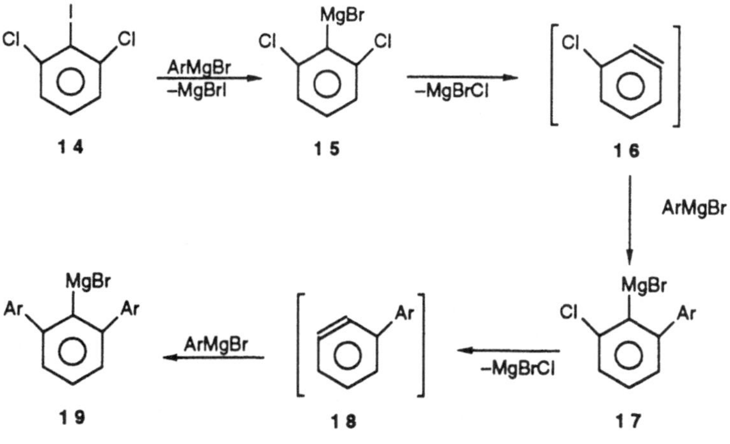

Scheme 1

$$(1)$$

$$(2)$$

MgBrCl generates aryne intermediate **16**. Regiospecific capture of **16** by a second equivalent of arylmagnesium bromide gives **17**, from which, for a second time the elements of MgBrCl are eliminated to yield aryne **18**. The third equivalent of arylmagnesium bromide regiospecifically captures **18** to yield the penultimate *m*-terphenyl intermediate **19**. Aqueous work-up of the reaction [Eq. (1)] gives the hydrocarbon **20** (E = H, D), but electrophilic quench of the reaction (Br$_2$, I$_2$, etc.) prior to aqueous work-up allows one to introduce a halo substituent on the internal (C$_{2'}$) position [6].

As evident from the mechanistic details (Scheme 1) a 'suicidal' equivalent of arylmagnesium bromide is needed to initiate the cascade of events. The consumption of an equivalent of arylmagnesium bromide just to effect the metal–halogen exchange is tolerable when the aryl Grignard reagent is either commercially available or easily procured. But in many a case it becomes advantageous to replace the 'suicidal' equivalent by a commercially available and cheap Grignard reagent [10, 14]. Vinylmagnesium bromide has been shown to effect the desired metal–halogen exchange [Eq. (2)] and is preferred over methylmagnesium bromide since the other product of exchange, namely vinyl iodide, is not susceptible to displacement by the Grignard.

A vast array of arylmagnesium bromides have been successfully employed for the synthesis of *m*-terphenyls by this route [6a]. In the present context of cuppedophane and cappedophane synthesis, the discussion can be limited to the use of a few arylmagnesium bromides (Scheme 2). The use of 2,6-dimethyl-phenylmagnesium bromide **21** followed by aqueous quench gave the *m*-ter-phenyl unit **7** in 70% yield [9]. The synthesis of **7** can easily be carried out on a 20 g scale. Electrophilic quench prior to aqueous work-up allows one to directly functionalize the internal position of the *m*-terphenyl. Detailed discussion of direct or indirect internal functionalization of *m*-terphenyl templates is given in Sect. 3.2.1.

Treatment of **14** with 3,5-dimethylphenylmagnesium bromide **25** gave the template **26** in 68% yield [11]. The *m*-terphenyl **26** and its internally function-alized derivatives serve as the molecular base for the general structure **4**. The use

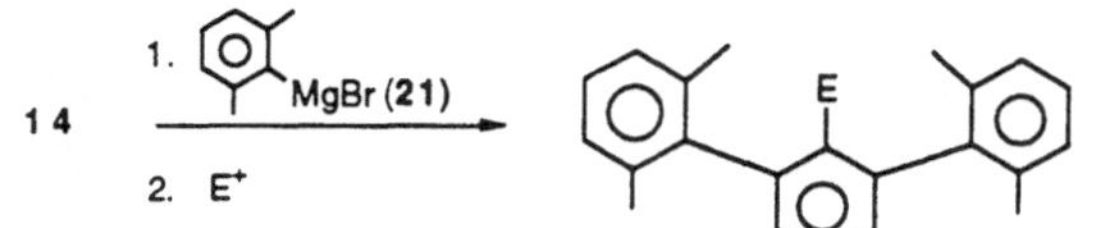

(3)

7 (E = H, 70%)
22 (E = D, 70%)
23 (E = Br, 62%)
24 (E = I, 82%)

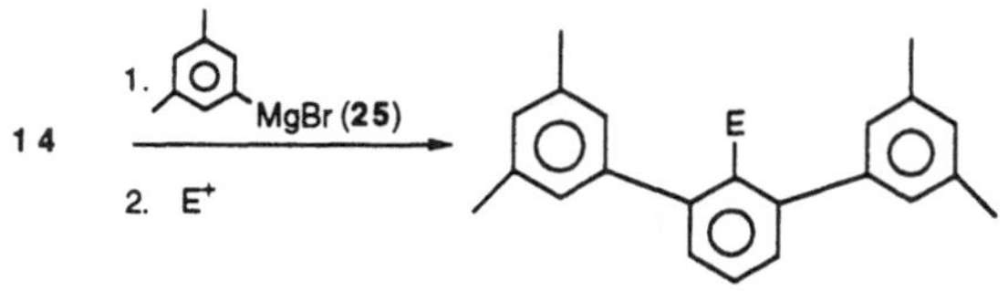

(4)

26 (E = H, 68%)
27 (E = D, 68%)
28 (E = I, 62%)

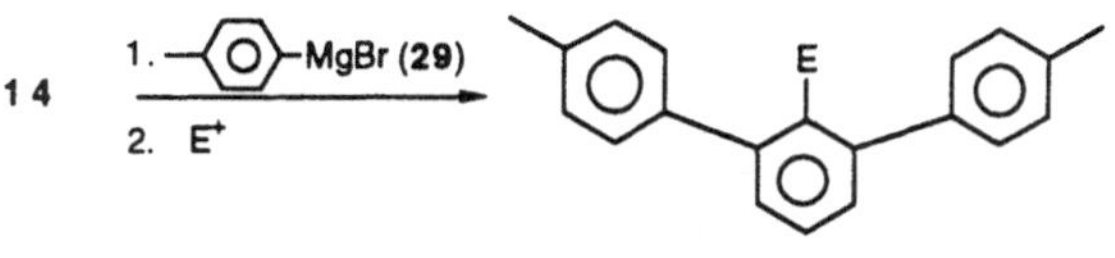

(5)

30 (E = H, 71%)
31 (E = D, 71%)
32 (E = Br, 75%)
33 (E = I, 60%)
34 (E = CO$_2$H, 55%)

Scheme 2

of 4-methylphenylmagnesium bromide 29 as the aryl Grignard provided a 71%
yield of the dimethyl-*m*-terphenyl 30 [15], the template for macrocycle 5.

The synthesis of *m*-terphenyls 7, 26 and 30 was also accomplished using
initial exchange by vinylmagnesium bromide followed by the use of only two
equivalents of the appropriate aryl Grignard. By this route, the *m*-terphenyls
were obtained in essentially the same yields as noted above, and with the added
advantage of not having the high boiling aryl iodide as a contaminant in the
product.

With 2,6-dimethoxyphenylmagnesium bromide 35 as the aryl Grignard one
obtains the tetramethoxyterphenyl 36 in 50% yield after aqueous quench

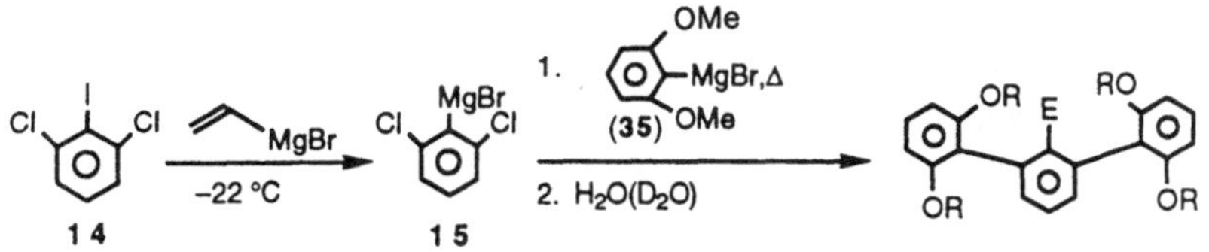

36 (E = H, R = Me)
37 (E = D, R = Me)

Scheme 3

125

[10, 16], or the deutero analogue **37** after D$_2$O quench. Since Grignard **35** is expensive, it is best to carry out the initial metal–halogen exchange with vinylmagnesium bromide (Scheme 3). Demethylated **36** (i.e., **13** (E = H) is an excellent template for cupped oxaphanes.

The 1,2,3-trihalobenzenes such as **14** (Scheme 1) may carry additional substituents on the ring, thus generating *m*-terphenyls with additional substituents on the central ring. Two examples are shown. Treatment of 2,4,6-tribromoiodobenzene **38** with three equivalents of **21** followed by aqueous work-up gave *m*-terphenyl **39** [17] with a C$_{5'}$ bromine substituent in 40% yield [Eq. (6)]. Similarly, the *m*-terphenyl with a C$_{4'}$ phenyl substituent **41** was obtained from 3,5-dibromo-4-iodobiphenyl **40** and **21** in 45% yield [Eq. (7)] [17]. *m*-Terphenyl-like structures with a 9,10-phenanthroline unit as the central ring

(6)

(7)

component (e.g. **47** and **48**) [18] were readily synthesized starting from phenanthroline and appropriate aryllithium reagents (Scheme 4). The reaction sequence involves either two successive Chichibabin additions followed by oxidation (Route a) or a pair of addition-oxidation steps (Route b). When the aryl unit was dimethoxyphenyl **47**, ether cleavage was possible using BBr$_3$ to give the tetraphenol **49** [18], used in the synthesis of several oxacuppedophanes.

3.2 Functionalization of *m*-Terphenyls

Functionalization of *m*-terphenyls within the context of the present review can be categorized into two types (i) introduction of a suitable functional group on the internal position of the central *m*-terphenyl ring; these functional groups are to be eventually encapsulated within the microenvironment of a cuppedophane

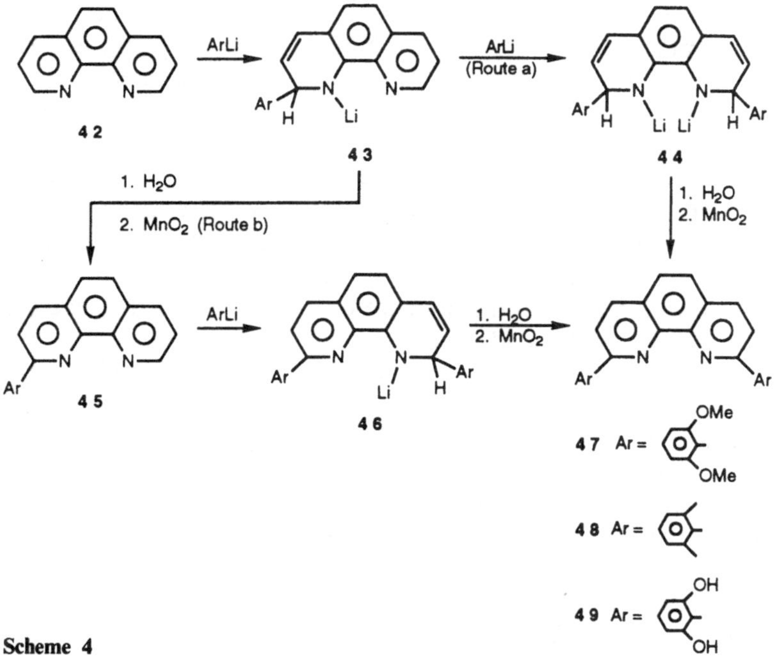

Scheme 4

or cappedophane, and (ii) introduction of functional groups on the outer *m*-terphenyl rings, subsequently to be employed for attaching tethers or capping units onto the base.

3.2.1 Functionalization of the C2' (Internal) Position on the Central Ring

The penultimate product of our *m*-terphenyl synthesis **19**, (Scheme 1), through reaction with electrophiles, allows for the direct introduction of a variety of functional groups at the internal position of the central *m*-terphenyl ring [6]. Halogens (Br or I) could be introduced by appropriate quench of the reaction prior to work-up. Even though the bromo derivatives **23**, **32** and the iodo derivatives **24**, **28**, and **33** (Scheme 2) were obtained in yields ranging from 60–82%, the direct introduction of other useful functionalities (formyl, carboxylic acid, etc.) was not possible in every case. The failure to introduce these carbonyl functionalities in acceptable yields by the direct quench method was pronounced in the tetramethyl-*m*-terphenyl cases [Eqs. (3) and (4)]. However,

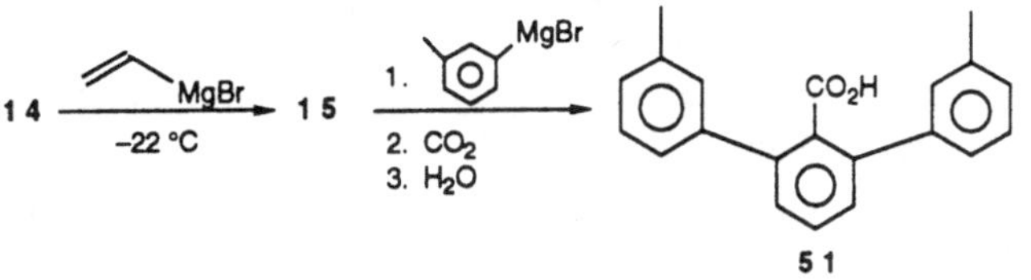

Scheme 5

127

the dimethyl derivative, 4,4"- dimethyl-1,1':3',1"-terphenyl-2'-carboxylic acid 34 (Scheme 2) was obtained in 55% yield by bubbling dry CO_2 through the reaction mixture prior to work-up [19]. Also prepared by the direct quench method is the 3,3'-dimethyl-1,1':3',1"-terphenyl-2'-carboxylic acid 51, which was obtained in 45% yield (Scheme 5).

The failure to incorporate versatile functionalities on the internal position of the central m-terphenyl ring led to the development of two alternate routes to procure such compounds. Lüning and co-workers successfully worked out a protocol for lithiation of readily prepared iodides 24 and 28 [20]. Treatment of iodo derivative 24 with n-BuLi in the presence of LiI in cyclohexane gave the lithio derivative 52 as a white solid which on subsequent treatment with dry CO_2 in THF gave a 71% yield of the carboxylic acid 53 (Scheme 6) [20]. Also prepared by the same procedure is the isomeric tetramethyl-m-terphenyl carboxylic acid 55 which was obtained in 50% yield. Both 53 and 55 could be esterified in almost quantitative yields by diazomethane to give 54 and 56 respectively. Saponification of the esters back to carboxylic acids could be affected in excellent yields using LiI in pyridine (Scheme 6).

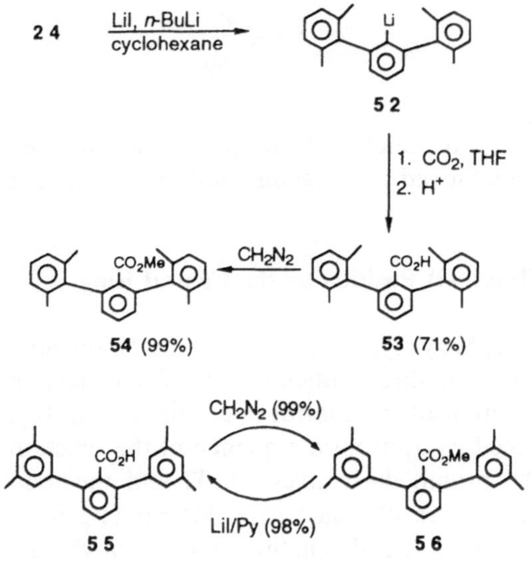

Scheme 6

The usefulness of the lithio derivative 52 as a precursor for the introduction of a variety of carbon and hetero atom derived functional groups was recently reported by Lüning and Baumgartner (Scheme 7) [21]. Reaction of 52 with DMF or formaldehyde gave the formyl (57, 61% [21]) and carbinol (58, 59% [21]) derivatives, respectively. 2'-Sulfur substituted m-terphenyls were synthesized using SO_2Cl_2 or SO_2 as electrophiles. Treatment of 52 with SO_2Cl_2 gave the sulfonyl chloride 59 [21] in 70% yield whereas SO_2 followed by acid work-up provided the sulfinic acid 60, also in 70% yield.

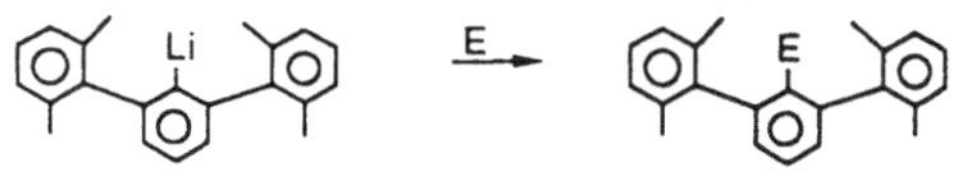

	E
5 7	CHO
5 8	CH₂OH
5 9	SO₂Cl
6 0	SO₂H
6 1	CH = CH₂
6 2	CH = NOH
6 3	CN
6 4	NC
6 5	SAc
6 6	SH

Scheme 7

The 2'-formyl derivative **57** was further subjected to functional group interconversions, leading to a series of a new 2'-substituted m-terphenyls. Wittig reaction on **57** provided the alkenyl derivative **61** in 58% yield [21]. The oxime **62** was obtained in 62% yield by treatment with hydroxylamine hydrochloride.

The sterically crowded nature of the $C_{2'}$ position is reflected in some of the reactivity patterns exhibited by functional groups occupying this unique position. For example, Beckmann rearrangement of oxime **62** was not possible with conc. H_2SO_4 and with PCl_5 gave a mixture of the cyanide **63** and the isocyanide **64** in 39% and 55% yields respectively. Arylsulfonyl chlorides are normally reduced to thiols by excess lithium aluminum hydride (LAH), and the intermediate sulfinic acids can only be isolated under well-defined reaction conditions. Sulfonyl chloride **59**, on the other hand, gave sulfinic acid **60** in 51% yield even after 15 h at reflux with a four-fold excess of LAH. The thiol **66** (64% yield) could only be accessed after two days of reflux with a 12-fold excess of LAH. A retarding influence of the two flanking dimethylphenyl groups was not always observed, however. For example, reduction of the sulfonyl chloride **59** using phosphorus and iodine in acetic acid/acetic anhydride provided an excellent 80% yield of the acetylated thiol derivatives **65**.

In our laboratories at Michigan State University, the approach to introducing useful carbonyl functionalities at $C_{2'}$ was to carry out, at the outset, a von Braun reaction using cuprous cyanide on iodide **24** (Scheme 8) [11]. Treatment of **24** (Scheme 2) with a 10-fold excess of CuCN in refluxing N-methylpyrrolidin-one (NMP) provided an 86% yield of the cyano derivative **63** [11]. Steric interference by the bulky dimethylphenyl groups prevented successful hydrolysis of this cyano group either to an acid or to an amide. However treatment of **63** with DiBAL-H in refluxing benzene transformed **63** to the 2'-formyl derivative **57** in 94% yield. Oxidation of **57** then provided an 86% yield of the acid **53**. Addition of methylmagnesium bromide to the 2'-formyl derivative in refluxing THF provided the carbinol **67** (86%), which upon pyridinium chlorochromate (PCC) oxidation provided the acetyl derivative **68** in 84% yield [11]. Our approach to introducing carbon substituents on $C_{2'}$ starting with a von Braun reaction (Scheme 8) is complementary to the one based on the lithio derivative **52** (Scheme 6) and offers the derivatives **53**, **57**, **67** and **68** in excellent yields.

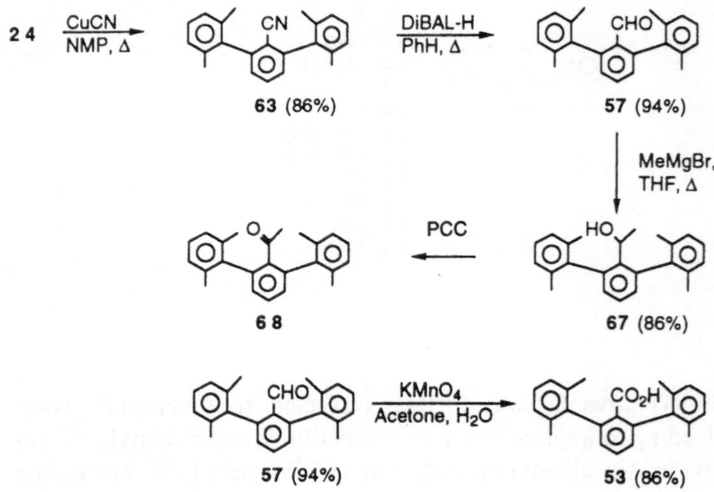

Scheme 8

3.2.2 Functionalization of the Outer *m*-Terphenyl Rings

Brief mention was made earlier (Sect. 2.2) to various types of functionalities on the outer *m*-terphenyl rings needed to assemble different cuppedophane and cappedophane structures.

2,6,2″,6″-Tetramethyl-1,1′:3′,1″-terphenyl 7 (E = X = H) was amenable to a selective four-fold monobromination of the methyl groups (Scheme 9). Treatment of 7 with the required amount of freshly recrystallized *N*-bromosuccinimide (NBS) (in portions) gave a respectable 62% yield of the tetrakis-bromomethyl compound 8 [9] which could be readily chromatographed to remove traces of polybrominated contaminants. Selective brominations such as this

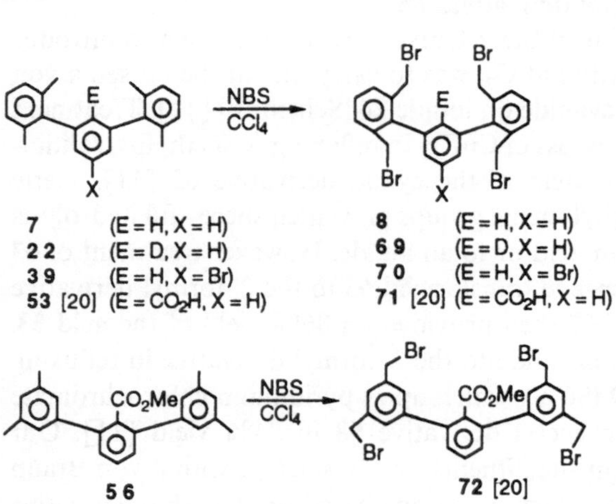

Scheme 9

were, however, a rarity. The only two other examples from our laboratory are the conversion of **22**, the deutero analog of **7**, to **69** [9] and the formation of **70** (E = H, X = Br [17]) from the corresponding tetramethyl compound **39** (E = H, X = Br). Lüning and co-workers have reported four-fold NBS bromination of the $C_{2'}$ substituted carboxylic acid **53** [20] as well as the $C_{2'}$-substituted carboxylic ester **56** [20]. The corresponding tetrakis-bromomethyl compounds **71** and **72** were obtained, after repeated recrystallizations, in acceptable purity for further reactions (vide infra).

The difficulty in effecting clean four-fold benzylic brominations, especially in tetramethyl-*m*-terphenyls bearing $C_{2'}$ functionality, prompted us to devise an alternate route to these molecules. The new route (Scheme 10) involves treatment of the tetramethyl compounds (**7, 23, 41, 63, 53**) with an excess of NBS to cleanly effect the transformation to the corresponding octabromides (**73–77**). The bromination proceeds in excellent yields and fortuitously only to the dibromo stage on each appendage, presumably, for steric reasons. Silver nitrate aided hydrolysis of the gem-dibromomethyl groups provided the tetraaldehydes (**78–81**) in excellent yields, although successful hydrolysis of the $C_{2'}$ carboxylic acid derivative **77** (to give **82**) could not be affected. Reduction of the tetraaldehydes to tetraalcohols (**9, 83–85**) proceeded in excellent yields. Subsequent treatment of these alcohols with PBr_3 in dry benzene provided the desired tetrakis-bromomethyl compounds (**8, 86–88**) in good yields. This route, though circuitous, provides tetrakis-bromomethyl derivatives of the highly desired internally functionalized *m*-terphenyls in good yields and free of polybrominated contaminants.

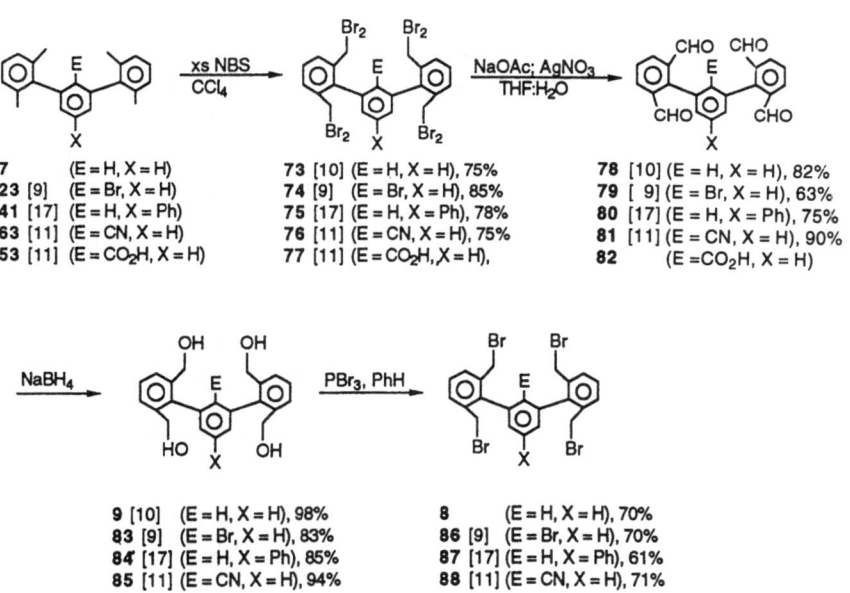

7	(E = H, X = H)
23 [9]	(E = Br, X = H)
41 [17]	(E = H, X = Ph)
63 [11]	(E = CN, X = H)
53 [11]	(E = CO₂H, X = H)

73 [10]	(E = H, X = H), 75%
74 [9]	(E = Br, X = H), 85%
75 [17]	(E = H, X = Ph), 78%
76 [11]	(E = CN, X = H), 75%
77 [11]	(E = CO₂H, X = H),

78 [10]	(E = H, X = H), 82%
79 [9]	(E = Br, X = H), 63%
80 [17]	(E = H, X = Ph), 75%
81 [11]	(E = CN, X = H), 90%
82	(E = CO₂H, X = H)

9 [10]	(E = H, X = H), 98%
83 [9]	(E = Br, X = H), 83%
84 [17]	(E = H, X = Ph), 85%
85 [11]	(E = CN, X = H), 94%

8	(E = H, X = H), 70%
86 [9]	(E = Br, X = H), 70%
87 [17]	(E = H, X = Ph), 61%
88 [11]	(E = CN, X = H), 71%

Scheme 10

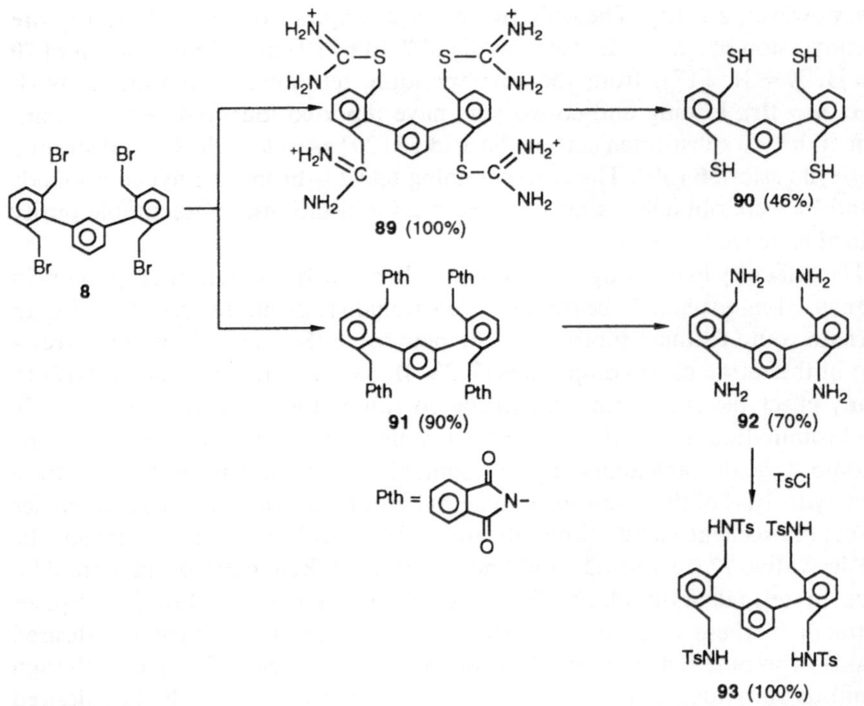

Scheme 11

The tetrakis-bromomethyl derivatives, apart from themselves being useful *m*-terphenyl base synthons for the construction of thia- and aza- cuppedophanes and cappedophanes (see Sects. 4.1 and 4.5), are also precursors to other useful molecular base synthons. Tetrakis-mercaptomethyl derivative **90** [9], tetrakis-aminomethyl derivatives **92** [11] and tetrakis-*p*-toluenesulfonylmethyl derivative **93** [11] are three new *m*-terphenyl templates easily accessed from **8** (Scheme 11). While the mercaptomethyl derivative was prepared via the isothiouronium salt **89**, the tetrakis-aminomethyl derivatives **92** and **93** were prepared via the phthalimido derivative **91** [11].

The various *m*-terphenyl base synthons described so far possess the common feature of having an intervening methylene unit between the *m*-terphenyl unit and the hetero atom functionality. The tetramethoxy *m*-terphenyls **36** and **37** (Scheme 3), on the other hand, are precursors with heteroatom functionality linked directly to the *m*-terphenyl unit. Demethylation of **36** and **37** provided the

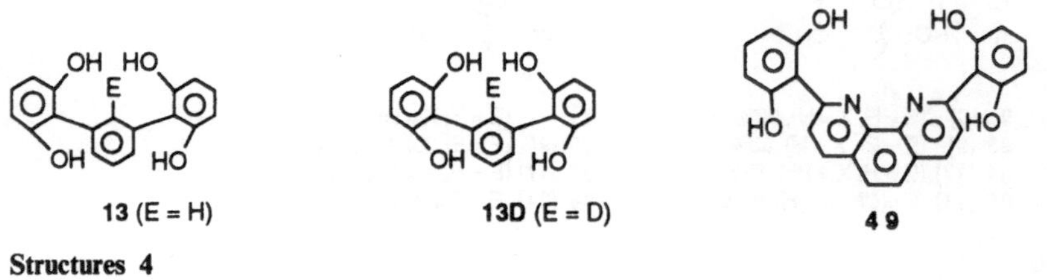

13 (E = H)   **13D** (E = D)   **49**

Structures 4

tetraphenols **13** and **13D** [10] in quantitative yields. The tetraphenols **13** and **13D** along with the tetraphenolic phenanthroline derivative **49** [18], provide base templates for cupped oxaphanes.

The tetrakis-bromomethyl compound **97**, a template for the construction of cuppedophanes of general structure **4**, was synthesized from the corresponding tetramethyl compound **26**. The circuitous 'octabromide' route (Scheme 10) was employed in this case, due to the inability to effect a clean, direct four-fold bromination of the methyl groups of **26**.

2 6   (X = CH₃, 68%)
94 [22] (X = CHBr₂)
95 [22] (X = CHO)
96 [22] (X = CH₂OH)
97 [22] (X = CH₂Br)

98 [15] (E = H, X = CH₂Br, 83%)
99 [15] (E = Br, X = CH₂Br, 78%)
100 [15] (E = I, X = CH₂Br, 87%)
101   (E = CO₂H, X = CH₂Br, 92%)
102   (E = H, X = CH₂SH, 86%)
103   (E = Br, X = CH₂SH, 68%)
104   (E = I, X = CH₂SH, 50%)
105   (E = CO₂H, X = CH₂SH, 38%)

Structures 5

Various templates, **98–105**, were readily synthesized from the corresponding internally substituted 4,4'-dimethyl-*m*-terphenyls and employed in the synthesis of macrocycles of general structure **5**. The bis-bromomethyl derivatives **98–101** [15] were obtained in good yields by the direct NBS bromination of the methyl groups. The conversion to the corresponding bis-mercaptomethyl compounds **102–105** [15] was carried out via the respective isothiouronium intermediates and were obtained in acceptable yields.

4 Cuppedophanes

4.1 Synthesis of Tetrathiacuppedophanes and NMR Evidence for the Cupped Structure

A widely used and often high-yield method for constructing cyclophanes is the high dilution coupling of thiols with reactive (usually benzylic) bromides [23]. The simplest of tetrakis-bromomethyl-*m*-terphenyls, namely, 2,6,2'',6''-tetrakis(bromomethyl)-1,1':3',1''-terphenyl (**8**) (Sect. 3.2.2) was initially chosen as the *m*-terphenyl template for the synthesis of tetrathiacuppedophanes. Xylylenedithiols were selected as the first choice for the tether (linking units) to assemble structures of general type **1** (Structures 1). The benzene rings of the dithiols were expected to confer rigidity to the 'cups' and deepen the molecular cavity by building up its walls. Coupling of *m*-xylylenedithiol **106** with bromide

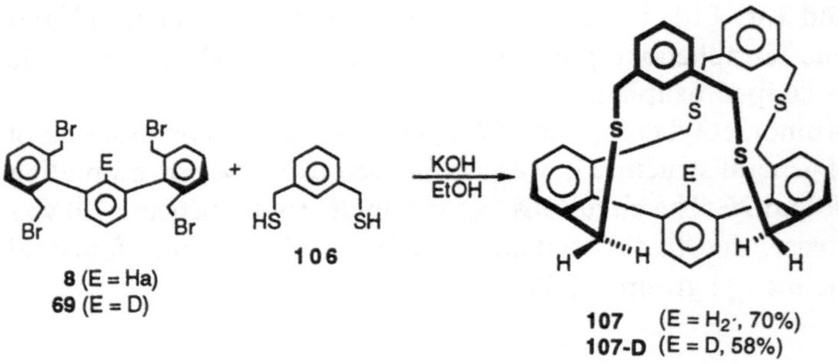

$$\text{(8)}$$

8 (E = Ha)
69 (E = D)

106

107 (E = H$_{2'}$, 70%)
107-D (E = D, 58%)

8, under high dilution conditions, gave a 70% yield of the tetrathiacuppedo-phane **107** [Eq. (8)] [9]. To assign a cupped structure (*m*-terphenyl unit being the base and the xylylene rings forming the walls) it was necessary to rule out alternate modes of coupling that could lead to structures **107a, 107b** and **107c**. In these alternate structures, the xylylenedithiols are coupled in a *syn*-fashion to each outer *m*-terphenyl ring (*anti*-conformations are unlikely for steric reasons). A reasonable model for **107a–c** is phenyl-[3,3]-*m*-cyclophane **108** [24], known (X-ray) to have the *syn*-conformation shown. The aromatic proton Hx in **108** appears at $\delta 5.47$ (20 °C). Should the coupling [Eq. (8)] have produced any or all of **107a–c** we should have seen either a two-proton signal or two one-proton signals at or near $\delta 5.5$. Instead, the cupped structure **107** shows, as its most diagnostic high field aromatic signal, a broad one-proton signal at $\delta 6.39$ assigned to H$_{2'}$, the isolated proton on the central *m*-terphenyl ring. This unique proton is shielded (relative to other aromatic protons in the molecule) by the two cofacially arranged *m*-xylylene units. The assignment was further confirmed by deuterium replacement, as in **107-D**. Also consistent with the cupped structure of **107** is a two-proton broadened singlet of $\delta 6.74$ assigned to the isolated protons on the *m*-xylylene linking units. These protons are in the shielding zone

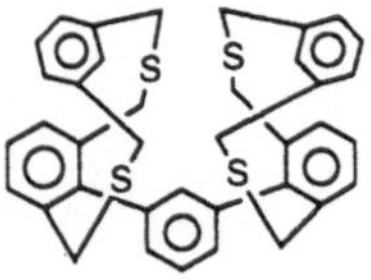

107a

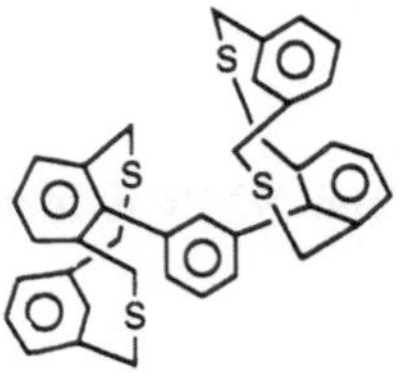

107b

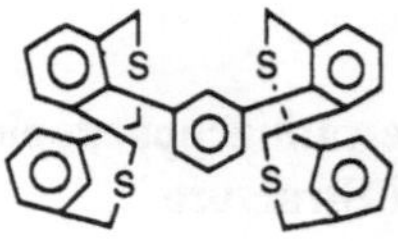

107c

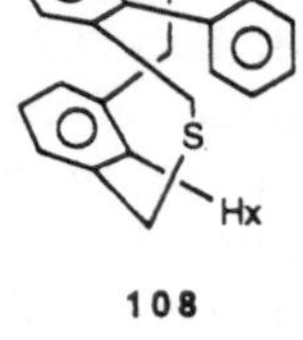

108

Structures 6

of the central ring of the *m*-terphenyl unit. The shielding effect of this central *m*-terphenyl ring is also reflected in the chemical shift value of the CH_2 unit nearest to the *m*-terphenyl base. The protons on the methylene unit appeared as two sets of doublets ($J = 11.1$ Hz) at $\delta 3.04$ and $\delta 3.36$ respectively, whereas the protons on the CH_2 unit nearest to the *m*-xylylene ring appeared as an ABq ($J = 14.1$ Hz) at $\delta 3.63$ and 3.51 respectively. The high field doublet at $\delta 3.04$ is assigned to the protons that point toward the central *m*-terphenyl ring (see structure **107**) and experience a shielding of about $\Delta \delta \approx 0.5$ ppm from the mean chemical shift value of an $ArCH_2S$ signal ($\delta \approx 3.5$) [25]. The rigid cupped arrangement of the rings in **107** is responsible for conferring this shielding on the CH_2 unit and the feature is observed in all cases. Section 5 compiles these data for a series of cupped tetrathiacyclophanes.

The use of 1,4-xylylenedithiol (**109**) and 1,2-xylylenedithiol (**110**) as the tether component in the coupling reaction [Eq. (8)] gave tetrathiacyclophanes **111** and **112** respectively. Structural details deduced from the NMR features of these products point to a cupped arrangement of rings in **111** but a flattened structure in **112** as drawn. The internal hydrogen ($H_{2'}$) of **111** appeared as a broad singlet at $\delta 6.26$, the assignment confirmed by the synthesis of **111-D**. The protons of the 1,4-xylylene units in **111** appeared as two broad singlets at $\delta 7.20$ and $\delta 6.83$, assigned respectively to the 'top' and 'bottom' protons on the rings (see structure). Heating a sample of **111** to 342 K in an NMR probe coalesced the broad singlets and at 378 K gave rise to a sharp one. This translates into $\Delta H^* = 17.0$ kcal/mol for the process (rotation about the 1,4-bond). This ΔH^* value safely precludes the alternate structure **111a** (or the two structures isomeric to **111a** where the 1,4-xylylene rings are either both on top or one on top and the

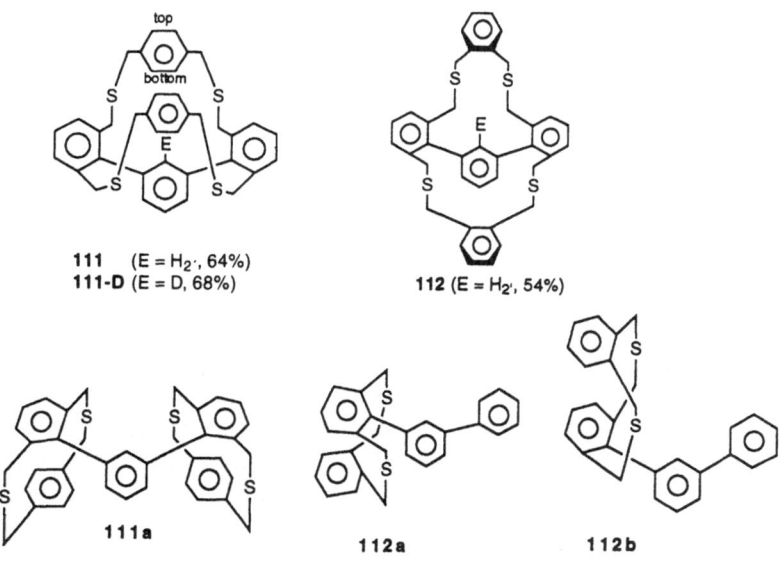

111 (E = H$_{2'}$, 64%)
111-D (E = D, 68%)

112 (E = H$_{2'}$, 54%)

111a

112a

112b

Structures 7

other at the bottom) for which CPK models show a very crowded structure in which the 1,4-xylylene unit severely impinges on the outer m-terphenyl rings. If the alternate structures were correct, we would also expect aryl protons on all four 'outer' rings to be highly shielded. The tetrathiacyclophane **112**, on the other hand, did not exhibit unique resonances for any of the aromatic protons, especially for the internal hydrogen $H_{2'}$.

The lack of features in the NMR spectrum of **112** and its non-resemblance to the isomeric structures **107** and **111** required the synthesis of a model compound in which the 1,2-xylylene unit is coupled across a single outer m-terphenyl ring. Comparison of NMR features with such a model compound would aid in assigning the mode of linkage of the tether (1,2-xylylene unit) to the m-terphenyl base. Model compounds **112a** and **112b** were synthesized as an inseparable mixture and the NMR features compared with that of **112**. The aromatic proton region of the NMR spectrum of **112a** and **112b**, as expected, was complex and of no particular comparative significance. However the methylene proton region was significantly different from that of **112**. In the tetrathiacyclophane **112**, the CH_2 unit nearest to the xylylene ring appeared as an ABq ($J = 11.8$ Hz) at $\delta 3.77$ and $\delta 3.49$ whereas the methylene unit nearest to the m-terphenyl base appeared as a broad singlet at $\delta 3.63$, not experiencing the shielding contribution from the central m-terphenyl ring. The corresponding resonances in the more rigid model compounds **112a** and **112b** appeared as four sets of AB quartets (of which 14 lines were visible) between $\delta 3.95$ and $\delta 3.00$. Although not unequivocal, these results strongly support a flattened structure for **112** as drawn.

The generality of the mode of linkage between the tethering unit and the m-terphenyl base was further evident from the formation of cupped tetrathiacyclophanes **115** [9] and **116** [17] [Eq. (9)]. The internal hydrogens of **115** and **116**

$$\text{(9)}$$

113 (R = OMe)
114 (R = OH)

115 (R = OMe, 73%)
116 (R = OH, 68%)

appeared at $\delta 6.35$ and $\delta 6.37$ respectively. The other significant resonances, almost identical in the two compounds and unique to a cupped arrangement of rings, are the following: the proton flanked by the $-CH_2S-$ groups on the m-xylylene unit appeared shielded at $\delta 6.33$ in **115** and at $\delta 6.29$ in **116**. The $-CH_2-$ protons nearest to the m-terphenyl base in both **115** and **116** appeared as two sets of doublets of which one set (at $\delta 2.46$) is markedly shielded by the central m-terphenyl ring. Such shielding, as mentioned above, is an indication of a rigid cupped arrangement of the rings (see also Sect. 5).

The locus of the thiol and benzylic halide functionalities can be reversed; thus tetrakis-mercaptomethyl-m-terphenyl (**90**) [9] can be coupled with bis-

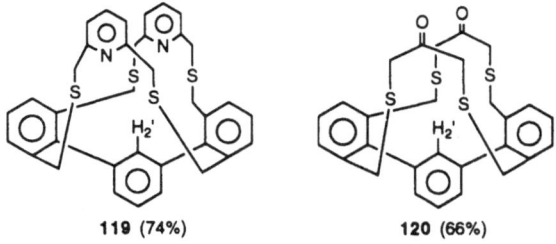

$$90 \quad + \quad 117 \xrightarrow[\text{Toluene-EtOH}]{\text{KOH}} \quad 118 \ (33\%) \tag{10}$$

benzylic halides leading to cupped tetrathiaphanes. Treatment of tetrathiol **90** with 9,10-bis(chloromethyl)anthracene (**117**) under high dilution conditions gave a 33% yield of the cuppedophane **118** [9], with a deep cavity [Eq. (10)]. The internal H$_{2'}$ of **118** appeared as a broad one-proton resonance at δ 5.03, indicating considerable shielding by the cofacial anthracene rings. The steric congestion exerted by the anthracene rings on the methylene units nearest to the *m*-terphenyl base probably forces their protons deeper into the shielding zone of the central *m*-terphenyl ring. This is reflected in the chemical shift value of the high field doublet of the –CH$_2$S– signal, which appeared at δ 1.61, i.e. considerably more shielded than a normal ArCH$_2$S signal ($\delta \approx 3.5$) [25].

Tetrathiacuppedophanes **119** and **120** (Structures 8) were also synthesized using **90** as the *m*-terphenyl base. Heterocuppedophane **119** has pyridine rings as wall components. The cupped arrangement of the rings is again evidenced by the chemical shift of the internal hydrogen (δ H$_{2'}$ = 6.54). The cupped tetrathiadiketone **120** [26], a desired precursor to a unique cappedophane (vide infra, Sect. 6.2), was prepared via a Cs$_2$CO$_3$ mediated coupling between **90** and 1,3-dichloroacetone. The cupped arrangement of the structure was proved by an X-ray structure (vide infra) and lends additional support to the generality that tethering occurs on the same side of the two outer rings rather than across a single ring.

119 (74%)   **120** (66%)

Structures 8

4.2 Intracavity Functionalized Tetrathiacuppedophanes

Macromolecules with sizable cavities having intraannular functional groups are of interest for at least two reasons. The intracavity functional groups in such compounds can provide stabilizing interactions (H-bonding, etc.) useful for trapping substrate molecules within the cavity. They also offer the opportunity to study the properties of those functional groups within a well-defined and controlled microenvironment. Macrocycles with intracavity functional groups can also be viewed as concave reagents [27] amenable to 'tailoring' to improve selectivity over similar reagents in which the same functional group lies outside a constrained environment.

Various $C_{2'}$ functionalized tetrathiacuppedophanes have been synthesized for these purposes. The tetrathiacuppedophane 121 [9], with a bromine atom at the $C_{2'}$ position, was synthesized by coupling pentabromide 86 (Scheme 10) with 1,3-xylylenedithiol (106) [Eq. (11)]. Although the viability of functional group

$$86 \;+\; 106 \quad \xrightarrow[\text{EtOH, PhH}]{\text{KOH}} \tag{11}$$

121 (E = Br, 67%)
122 (E = Li)

interchange (via a metalation and appropriate electrophilic quench) in 121 made this molecule an attractive precursor to various $C_{2'}$-functionalized cuppedophanes, in practice, the desired interconversions proved to be unattainable. The lithio derivative 122 was readily formed through a metal–halogen exchange with butyllithium on 121, but it could only be successful quenched with H_3O^+ (generating the cuppedophane 107) and failed to be captured by other electrophiles such as CO_2, DMF and N-formylpiperidine. The inability to employ 121 to introduce useful functionalities at the $C_{2'}$ position prompted us, at Michigan State, to synthesize cuppedophane 123 [11] with a cyano group at the internal position. Although 123 was easily prepared [Eq. (12)], attempted functional

$$88 \;+\; 106 \quad \xrightarrow[\text{EtOH, PhH}]{\text{KOH}} \tag{12}$$

123

group interconversions such as $CN \rightarrow CHO$ using DIBAL-H as well as $CN \rightarrow C(O)Me$ using MeMgBr failed, presumably because of steric inaccessibility to the reaction center (cf. Scheme 8).

The synthesis and successful purification of $C_{2'}$-COOH/COOR substituted tetrakis-(bromomethyl)-m-terphenyls 71 and 72 (Scheme 9) aided Lüning and

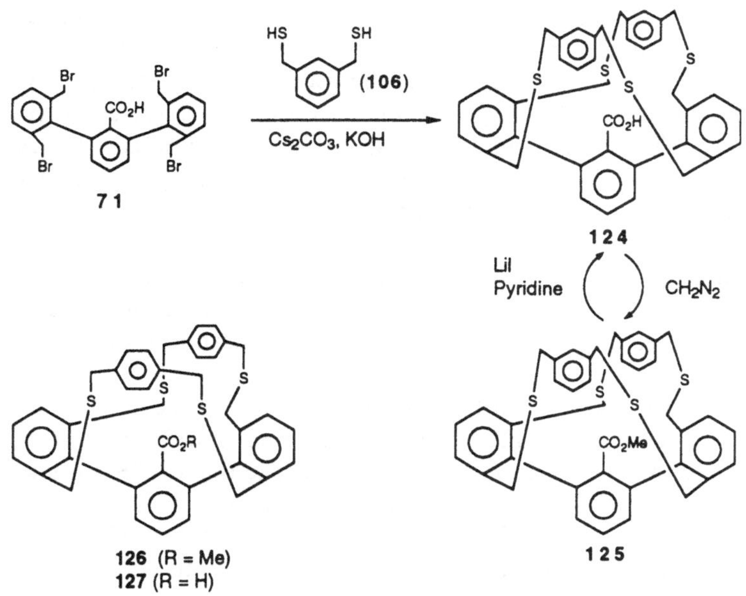

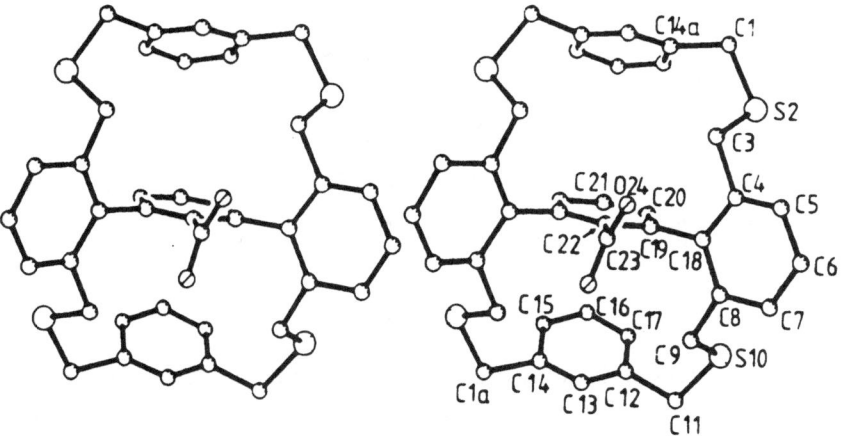

Scheme 12

coworkers [20] to synthesize the first cuppedophanes with useful and interesting functionalities. Treatment of the bromotemplate **71** with 1,3-xylylenedithiol (**106**) under high dilution conditions in the presence of KOH–Cs$_2$CO$_3$ provided the first cupped tetrathiaphane with a carboxyl functional group inside (Scheme 12). Purification was greatly aided by directly converting acid **124** to its methyl ester **125** [20] with diazomethane. Regeneration of the acid from the ester required prolonged heating with lithium iodide in pyridine [27] indicating the

Fig. 1. Stereo plot of the bimacrocyclic acid **124** as determined by X-ray analysis

steric inaccessibility to the reaction center. A three dimensional structure of **124** (Fig. 1) revealed that the carboxyl group is embedded between the outer *m*-terphenyl rings. Based on the X-ray data, the accessibility of the carboxyl group was investigated by calculating the solvent-accessible surface using the Connolly routine [28]. (In the Connolly routine, spheres of various sizes are rolled over the van der Waals surface of the molecule and the resulting contact surface monitored). The calculations showed that the carboxyl oxygen atoms and the hydrogen atoms of the outer *m*-terphenyl rings of **124** approximately form a plane, indicating a topology borderline between concave and convex for the molecule. The *m*-xylylene rings of **124** are folded away from the carboxyl functionality (see Fig. 1), thereby 'exposing' the carboxyl functional group.

Exchange of the *m*-xylylene tether by a *p*-xylylene one led to the much stiffer tetrathiacuppedophanes **126** and **127** [20] with fairly well-embedded carboxyl groups. The synthesis and purification of **127** was similar to that of **125**. Various tetrathiacuppedophanes with a larger degree of concave shielding were also recently synthesized by Lüning and coworkers [29]. Tetrathiacuppedophanes **128** and **129** [29] (Structures 9) have anthracene units as wall components. These anthracene units make the structure much more rigid (and consequently more concave) and also increase the depth of the cavity. The anthracene-incorporated tetrathiacuppedophanes **118** [Eq. (10)], **128** and **129** are fairly insoluble compounds, but the *tert*-butyl groups in **128** and **129** help to increase their solubility. In **130** and **131** (Structures 9) the tethers (based on 2,7-dimethylnaphthalene) are linked to the 3,3″ and 5,5″ positions on the outer *m*-terphenyl rings (cf. general structure **3**). Such an arrangement increases the concavity of the structure with the depth of the cup taken care of by the naphthalene wall components.

128 (R = Me)
129 (R = H)

130 (R = Me)
131 (R = H)

Structures 9

The relative extent of shielding of the ester functionality in tetrathiacuppedophanes **125**, **126** (Scheme 12) and **128**, **130** is evident from the reaction conditions required to effect ester cleavage in these molecules [29]. Whereas the ester functionality in **125** and **126** could be cleaved using LiI in pyridine at reflux, the ester function in **130** required the use of crown ethers to increase the nucleophilicity of iodide ion in pyridine. The highly shielded ester group in cuppedophane **128** required LiI and the higher temperatures provided by 4-methylpyridine as the solvent to effect the same conversion.

Another functionality introduced into the concave base of a tetrathiacuppedophane is the thioacetyl group [21]. The tetramethyl-*m*-terphenyl **65**, with a $C_{2'}$ thioacetyl group, was successfully tetra-mono-brominated to yield **132** [21], which was subsequently coupled with 1,3-xylylene dithiol to yield **133** (Scheme 13), albeit in very low yield (2%). The structure of **133** is tentatively drawn with the xylylene units folded away from the thioacetyl functionality, similar to the solid state structure (Fig. 1) of **124** [20]. The chemical shift reported for the thioester methyl group of **133** ($\delta 2.50$) is in fact about 0.7 ppm deshielded from the corresponding signal in **65** ($\delta 1.80$). This discrepancy, along with the deshielded nature of the $-CH_2S-$ nearest to the *m*-terphenyl base (see Sect. 5) suggests the need for an X-ray structure before a cupped conformation can be assigned to **133**.

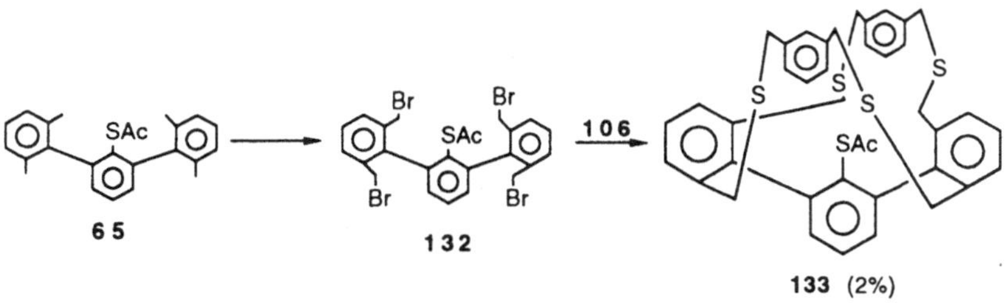

Scheme 13

The extent of enclosure of the ester functionality by the wall components of tetrathiacuppedophanes can be judged by the diamagnetic shielding experienced by the methyl groups. Table 1 lists chemical shift values of the methyl protons of the methyl ester in various tetrathiacuppedophanes. Included in parenthesis is the extent of shielding, i.e. $\Delta\delta = \delta$ ester (cuppedophane)$-\delta$ ester (*m*-terphenyl).

m-Terphenyls synthesized as described in Sect. 3.1, where various groups are easily placed at $C_{2'}$, are ideal precursors of macrocyclic phanes bearing intracavity functionality. Though not strictly cuppedophanes, these easily assembled

Table 1. Chemical shift value of methyl ester protons in various tetrathiacuppedophanes

Tetrathiacuppedophane	ppm ($\Delta\delta$)
125	2.88 (− 0.20)
126	2.75 (− 0.33)
128	2.21 (− 0.84)
130	1.94 (− 1.48)
133	2.50 (+ 0.7)

large-ring phanes may find similar utility in binding guests within their cavities. Only a few representative examples are presented here [19].

Condensation of bis-bromide **134** with bis-thiol **135** (cf. **98–105**), or coupling of bis-bromide **134** with sodium sulfide, leads in acceptable yields to macrocycles **136** [Eq. (13)]. Examples of E_1 and E_2 include H, D, Br, I, CN, CO_2H, CO_2R. Larger phanes with intracavity functionality can be obtained, for example, by coupling two equivalents of bis-bromide **134** with o-, m- or p-xylylene dithiols. The methodology is very versatile, since 3,3″-analogues of **134** and **135** are also readily prepared and coupled.

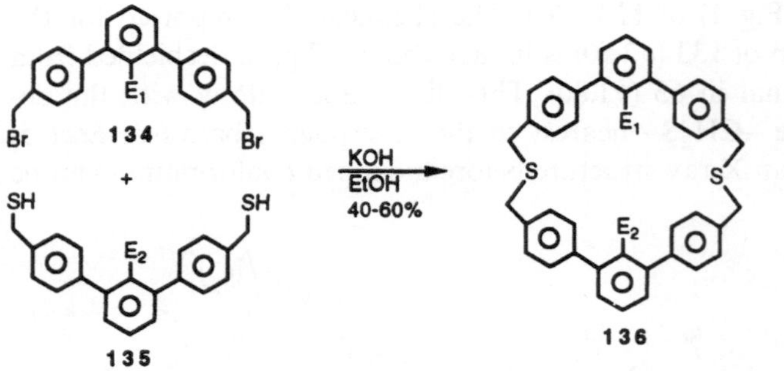

$$(13)$$

The m-terphenyl moieties can first be linked at E_1 and E_2, then converted to 'bicyclic' phanes, as shown in Scheme 14. In the final step, the outer rings of the two m-terphenyl moieties of tetrabromide **139** can also be linked by xylylene dithiols. Related compounds such as **141** have recently been prepared [30] as potential amino acid receptors.

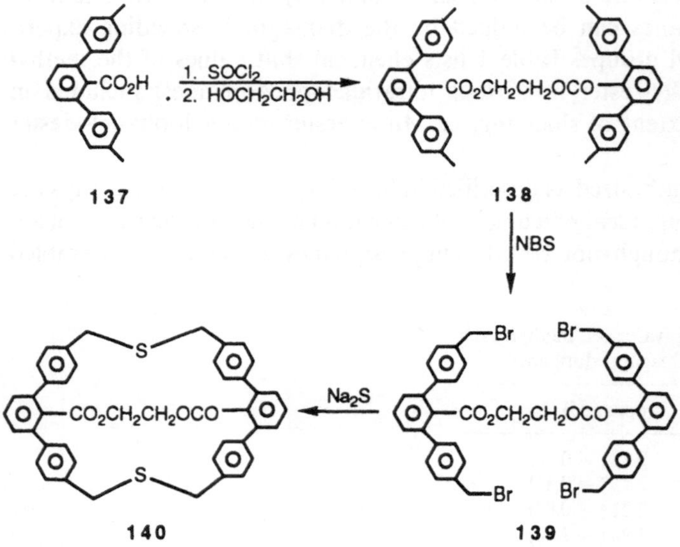

Scheme 14

141

Structure 10

4.3 Tetraoxacuppedophanes and Analogues with Concave Basic Sites

m-Terphenyls in which the outer rings bear phenolic (**13** and **13-D**) or hydroxy-methyl groups (**9** and **9-D**) [10] were mentioned earlier as potential templates for oxacuppedophanes. The first of these to be used as the framework for an oxacyclophane were the tetraphenols **13** and **13-D**. Coupling of **13** with 1,3-xylylene dichloride (**142**) in DMF in the presence of K_2CO_3 provided a 34% yield of oxacuppedophane **143** [Eq. (14)] [10]. The cupped arrangement of rings is

$$(14)$$

143 (34%)

apparent from the diamagnetic shielding experienced by $H_{2'}$, which appeared as the highest field aromatic resonance in its NMR spectrum, at $\delta\,6.60$. Though one might ascribe the cause of this shielding solely to a diamagnetic contribution of the two cofacial *m*-xylylene rings, it is possible that in strained cuppedophanes like **143** (where tethers are linked directly to the outer rings) $H_{2'}$ could very well be in the combined shielding zones of four aromatic rings, i.e. two *m*-xylylene rings and two outer *m*-terphenyl rings. This hypothesis is explained with the aid of Fig. 2 and is substantiated below with additional examples. The consequence of shortening the tethers by directly attaching them to the outer *m*-terphenyl rings is a severe steric compression between the central *m*-terphenyl ring and the two cofacial rings that are part of the tethers. One way to relieve such a strain would be to bend the two outer *m*-terphenyl rings toward each other (by an angle θ as shown in Fig. 2). This would raise the *m*-xylylene rings of the tethers, removing their colinearity with the central *m*-terphenyl ring. Protons H_a are thus removed from the shielding zone of the central *m*-terphenyl ring and appear

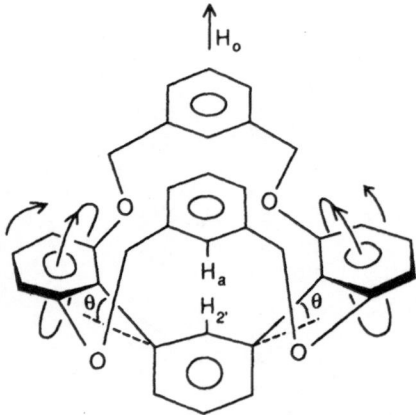

Fig. 2. Shielding of $H_{2'}$ in strained tetraoxacuppedophanes

at $\delta 7.43$, vis-a-vis $\delta 6.74$ in **107** [9]. An additional diagnostic consequence of tilting the outer *m*-terphenyl rings is the diamagnetic shielding exerted by these rings (from their new position) on $H_{2'}$ which appears at $\delta 6.60$, vis-a-vis $\delta 6.39$ in **107**. A pronounced manifestation of the distortion described above is found in tetraoxacuppedophane **145** [10] where the two methoxyl groups on the *m*-xylylene rings increase the steric compression with the central *m*-terphenyl rings. High dilution coupling of tetraphenol **13** with 2,6-bis(bromomethyl) anisole (**144**) gave two coupled products, **145** and **146** [10] in 9% and 27% yields respectively. These products are assigned the up-up and up-down conformations (as drawn, Scheme 15) for reasons explained below. The descriptions up-up

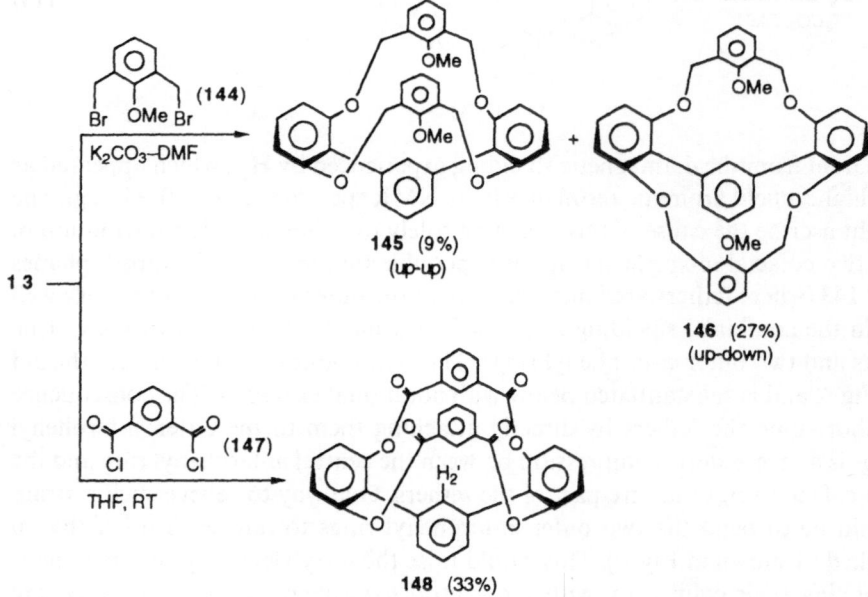

Scheme 15

and up-down refer to the orientation of the anisyl rings relative to the central *m*-terphenyl ring. The presence of two anisyl rings in **145** makes the up-up conformer sterically more demanding than the analogous structure **143** (devoid of methoxyl groups) and as postulated experiences a greater deviation of the outer *m*-terphenyl rings towards each other. This is reflected in the chemical shift of $H_{2'}$, which appeared at $\delta 3.84$ (vis-a-vis $\delta 6.60$ in **143**), an assignment verified by the synthesis of the deutero analogue from **13-D**. The chemical shift of $H_{2'}$ in **146** is $\delta 5.70$, between that of the corresponding protons in **143** and **145**, as expected. The relative orientation of methoxyl groups in **145** and **146** were decided by NOE methods. The oxaphane **146** with two methoxyl resonances ($\delta 3.24$ and $\delta 3.56$) clearly required an up-down arrangement for the anisyl rings. While irradiation of the methoxyl signal at $\delta 3.56$ caused a 5% intensity enhancement in the resonance of $H_{2'}$ ($\delta 5.70$), irradiation of the second methoxyl signal caused no change in $H_{2'}$ intensity. A CPK model of **146** indicated that the methoxyl on the 'down' ring is closer to $H_{2'}$ allowing an assignment of the resonance at $\delta 3.56$ to the 'down' methoxyl. The similarity in the chemical shift of the remaining methoxyl ($\delta 3.24$) in **146** and the methoxyl resonance in **145** ($\delta 3.44$) confirms the up-up arrangement of rings in **145**.

The tetraphenol template **13** has also been used to introduce acyl tethers. Treatment of **13** with isophthaloyl chloride (**147**) in the presence of triethylamine gave a 33% yield of the cyclic ester **148** [10] (Scheme 15). Replacement of the CH_2 groups in **143** with carbonyl groups widens the tethers and consequently demands no or considerably less distortion of the outer *m*-terphenyl rings. This is reflected in the chemical shift of $H_{2'}$ which appeared at $\delta 7.00$ in the NMR spectrum of **148**. Even though the appearance of the $H_{2'}$ resonance at $\delta 7.00$ as the most upfield aromatic resonance in **148** implies a cupped arrangement of the rings, the extent of shielding in considerably less than what is observed in other cuppedophanes, for examples $\delta 6.37$ in **107**, $\delta 6.26$ in **111** and $\delta 6.35$ in **115**. For this reason, an X-ray structure determination of **148** should be done before a conformational assignment is assumed.

Tetrakis(hydroxymethyl)-*m*-terphenyl **9** [10] and its deutero analog **9-D** have also been successfully employed as templates for oxacuppedophanes. Treatment of tetraalcohol **9** with 1,3-xylylene dichloride (**142**) in DMF in the presence of NaH gave a 42% yield of tetraoxacuppedophane **149** [10] [Eq. (15)]. The appearance of the $H_{2'}$ resonance at $\delta 6.71$ (verified by the synthesis of **149-D**)

$$(15)$$

9 (E = H)
9-D (E = D)

149 (E = H)
149-D (E = D)

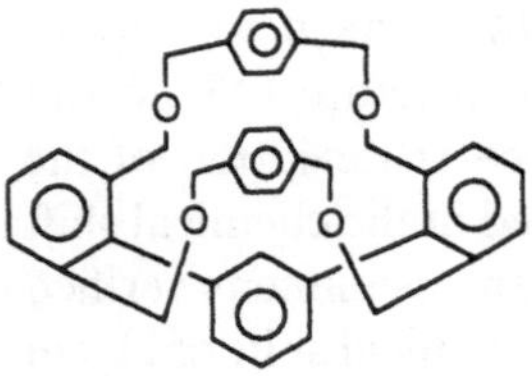

150 (35%)

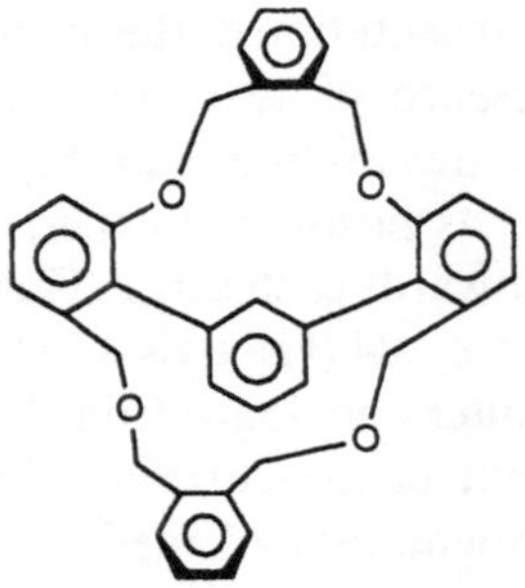

151 (20%)

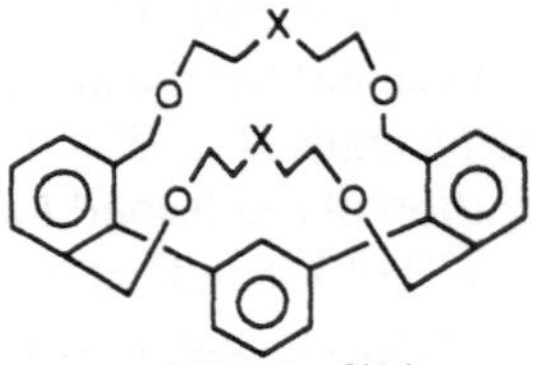

152 (X = CH₂)
153 (X = O)

Scheme 16

confirms a cupped arrangement of rings in **149**. The *p*-xylylene analog **150** (Scheme 16) also has a cupped arrangement of rings, as evidenced by the H$_{2'}$ resonance at $\delta 6.21$. The *o*-xylylene analog **151** (Scheme 16) has a flattened structure deduced for reasons similar to those for the flattened tetrathiacuppedophane **112**. Aliphatically linked oxacyclophanes **152** and **153** [10] (Scheme 16) were synthesized for comparison with the aryl-bridged cyclophanes. The yields of **152** and **153** were low (15–16%) and as expected, there were no unique aromatic resonances in these structures.

The use of the phenanthroline template **49** (Scheme 4) with the outer rings bearing four phenolic groups generates tetraoxacuppedophanes with concave basic functionality [8, 18]. Treatment of tetraphenol **49** with two equivalents of triethyleneglycol ditosylate (**154**) in the presence of NaH in DMF under high dilution conditions gave a 13% yield of **156** [18] [Eq. (16)]. An isomeric structure possible from this coupling reaction is the metacyclophane structure **158**, where the ethers have bridged across the two phenolic groups of the same

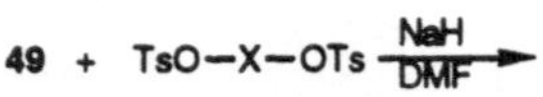

49 + TsO—X—OTs $\xrightarrow[\text{DMF}]{\text{NaH}}$

154 [X = CH₂(CH₂OCH₂)₂CH₂]
155 [X = CH₂(CH₂OCH₂)₃CH₂]

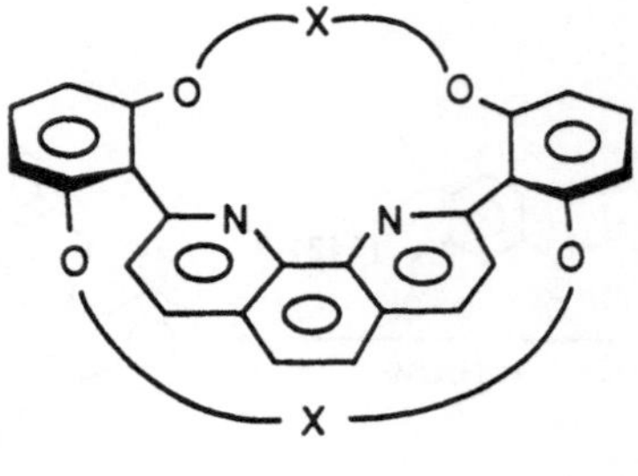

(16)

156 [X = CH₂(CH₂OCH₂)₂CH₂)]
157 [X = CH₂(CH₂OCH₂)₃CH₂)]

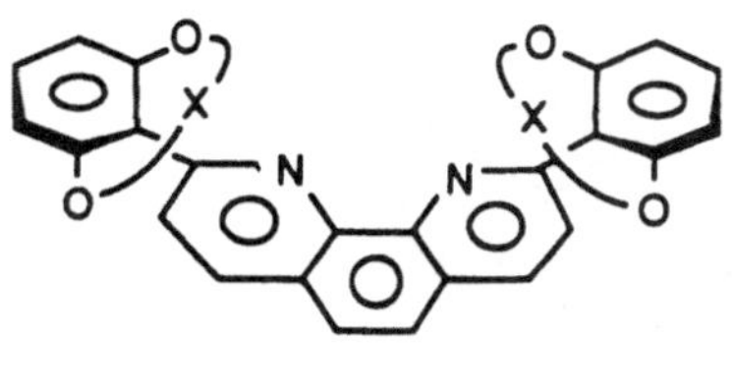

158

Scheme 17

ring. NOE measurements indicated that irradiation of the central dimethylene unit on the bridge had no effect on the aromatic protons of the outer rings of the molecular base. A metacyclophane structure such as **158** would be expected to have aromatic protons well within the proximity of the irradiated dimethylene unit to experience NOE enhancement. Additional evidence for this mode of tethering was observed when the NMR spectrum of the complex between **156** and picric acid showed an upfield shift for all eight hydrogens of the central dimethylene unit. The cause of the upfield shift is believed to be the close proximity of the central dimethylene unit to the protonated nitrogens of the phenanthroline ring. The tetraoxaphane **157**, with tetraethyleneglycol units as the tethers, was also similarly prepared [Eq. (16)].

Additional examples of tetraoxacuppedophanes based on the phenanthroline template are **159** and **160** with polymethylene chains as the bridging units, and the chiral concave base **161** [8].

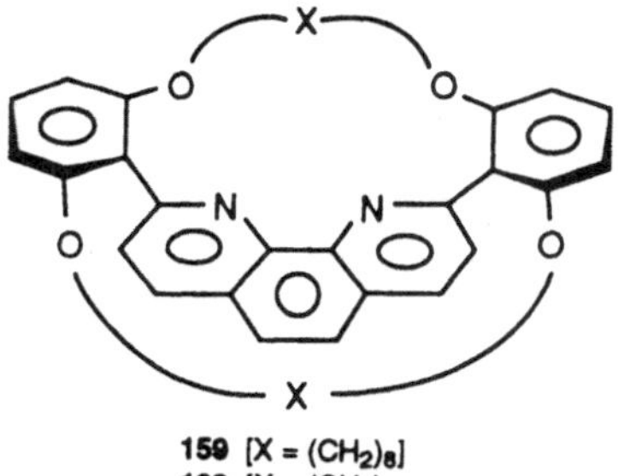

159 [X = (CH$_2$)$_8$]
160 [X = (CH$_2$)$_{10}$]

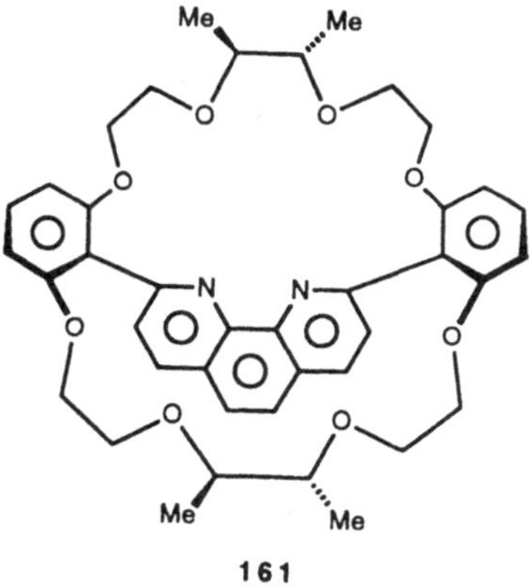

161

Structures 11

4.4 1,10-Phenanthroline-Based Cuppedophanes as Selective Concave Reagents

Selectivity in enzymatic reactions is in part due to the fact that the catalytic sites are buried in concave cavities so that only substrates with optimum fit are acted upon. The ease of modifying cavity size and the presence of catalytic sites (either basic N or its protonated form, $^+$NH) within the concave cavity makes

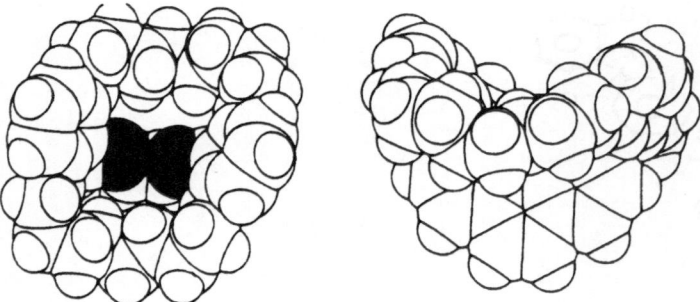

Fig. 3. Two views of the tetraoxophane **159**. (The *darkened atoms* are the phenanthroline nitrogens)

macrocycles such as **156** [18] an excellent choice for testing their efficiency as concave reagents for model reactions. Figure 3 shows two views of the concave phenanthroline-based tetraoxaphane **159**. Accessibility of the nitrogen atoms of **159** was measured using the Connolly routine [28]. It was found that molecules or parts of molecules with a diameter larger than 5.6 Å will not be able to come into contact with the basic nitrogen atoms. The model reaction that was selected to test the efficiency/selectivity of these reagents was the protonation of nitronate anions [31].

4-Phenylcyclohexyl nitronate **162** has three protonation sites (Scheme 18). Protonation at oxygen results in ketone **163** after hydrolysis, whereas *cis* or

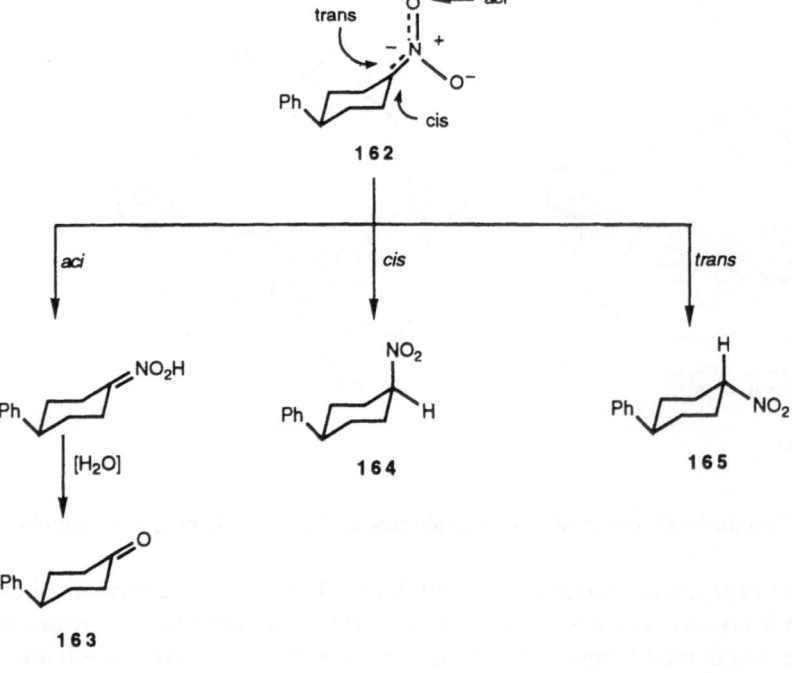

Scheme 18

$$EtOH_2^+ \xrightarrow[H_2O]{RR'C^{-}-NO_2} RR'C = O$$

$$\xrightarrow{H^+} \xrightarrow[\text{(C Protonation hindered)}]{RR'C^{-}-NO_2} RR'CH-NO_2$$

Scheme 19

trans protonation of the α-carbon atom leads to stereoisomeric nitro compounds **164** and **165**. The Nef reaction (O-protonation of the nitronate anion followed by hydrolysis) is usually carried out in fairly acidic media. However if C-protonation is hindered by using a protonated concave reagent, the Nef reaction will be effected by the conjugate acid of the solvent. Lüning and coworkers, who have studied such reactions, suggest that Nef reactions in weakly acidic media be designated the "soft-Nef" reaction. The explanation offered for the chemoselectivity in protonation is as follows: In concave buffers the protons within the cavity are relatively inaccessible for C-protonation; consequently the Nef reaction proceeds via O-protonation by $EtOH_2^+$ (Scheme 19). For example, concave pyridine macrocycle **166** [32] exhibited a chemoselectivity for O-protonation of nitronate anion in a buffered medium.

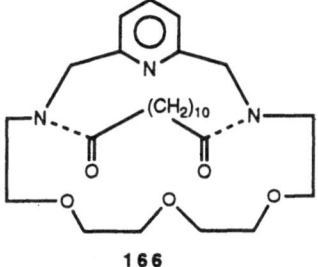

166

Structure 12

Macrocycles of this type are related to but do not fall precisely under the category of cuppedophanes, and readers are referred to the original papers [31, 32] for details. Neither chemoselective O-protonation of the nitronate anion **162** nor stereoselective C-protonation was observed when cuppedophanes **156, 157,** or **159** were used as the concave reagent in a buffered medium. However, with 2-[bis(*ortho*-substituted)aryl]1,10-phenanthrolines **167–173** [8] in a buffered medium, the thermodynamically less stable *cis*-4-phenylnitrocyclohexane **164** was formed in preference to the *trans*-isomer **165** (Scheme 20). The reason for the selectivity is thought to be the trapping of the proton by the concave reagent, thus making it inaccessible. Approach of the intra-cavity proton from the "top" leading to transition state **165–T.S.** is sterically prohibitive (severe 1,3-diaxial interactions); hence the thermodynamically less stable **164** is produced via

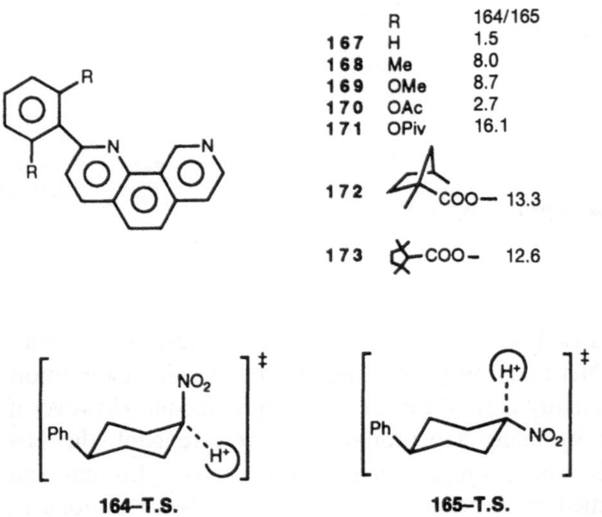

	R	164/165
167	H	1.5
168	Me	8.0
169	OMe	8.7
170	OAc	2.7
171	OPiv	16.1

172   13.3

173   12.6

164–T.S.   **165–T.S.**

Scheme 20

transition state **164–T.S.** It is not altogether clear why similar selectivity is not observed with macrocycles **156**, **157** and **159**, but one possibility is that the tethers sterically prevent the carbanionic site from approaching the intracavity proton.

While stoichiometric amounts of base were used in the nitronate anion protonations, catalytic amounts of concave macrocycles **156**, **157** and **159** were found to influence the addition of alcohols to diphenylketene. The chiral concave macrocycle **161** catalyzed the addition of R-1-phenylethanol 20% faster than the addition of the S-enantiomer to the ketene thus demonstrating the enantioselectivity of the concave reagent [8].

4.5 Tetraazacuppedophanes, and Conformational Aspects of Their N-Tosylamide Precursors

Cuppedophanes of general structure **1** with N-atoms linking the m-terphenyl base and the wall components have been reported [33]. Such tetraazacuppedophanes were accessed via the corresponding N-tosylamides [34]. m-Terphenyl base templates employed in these syntheses include the tetrakis(bromomethyl)-m-terphenyl, **8** [9] (Scheme 9) and the tetrakis[(p-tolylsulfonyl aminomethyl)]-m-terphenyl **93** [11] (Scheme 11). Coupling of **8** with 1,3-bis[p-tolylsulfonyl aminomethyl)]benzene (**174**) in DMF in the presence of cesium carbonate at RT gave the cupped tetrakis-N-tosylamide **175** [33] in 74% yield (Scheme 21). The cupped tetrakis-N-tosylamide **176** (41%) and the flattened (as drawn) macrocyclic tetrakis-N-tosylamide **177** (60%) (Scheme 21) were similarly prepared from **8** and the appropriate xylyleneamine ditosylate [33]. Evidence for the cupped (**175** and **176**) and the flattened (**177**) structures is discussed below.

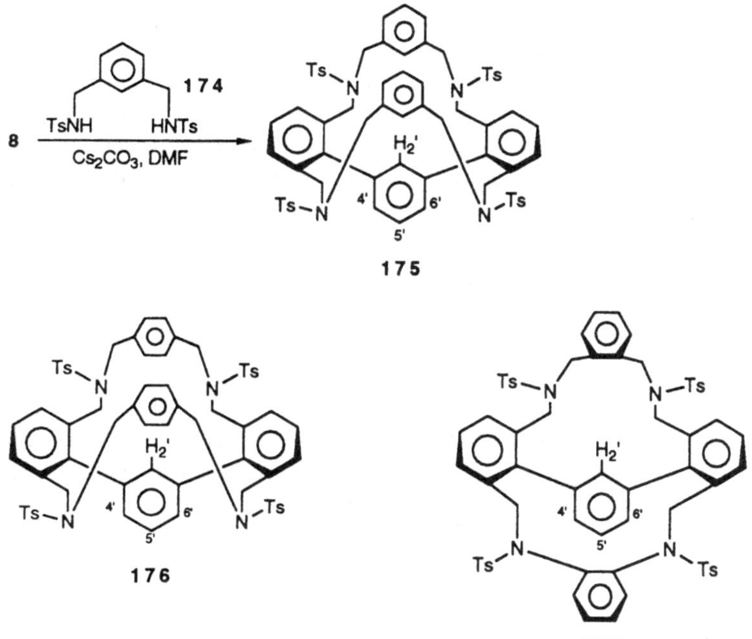

Scheme 21

(17)

The internal proton (H$_{2'}$) of **175** appeared as a broad singlet at $\delta 4.56$, considerably more shielded than the corresponding proton in the S-analogue **107** ($\delta 6.39$) and the O-analogue **149** ($\delta 6.71$). An even more striking difference was found in the chemical shift of the internal proton (H$_{2'}$) of **177**, which appeared at $\delta 4.76$ whereas the corresponding proton (H$_{2'}$) in the flattened analogues **112** (thia) and **151** (oxa) appeared between $\delta 7.30$–7.10. Unlike in **175** and **177** the internal proton (H$_{2'}$) of the *p*-xylylene linked tetrakis-*N*-tosylamide **176** has a chemical shift value of $\delta 6.26$, almost identical to that of the corresponding proton in the S-analogue **111** ($\delta 6.26$) [9] and the O-analogue **150** ($\delta 6.21$) [10]. These anomalous chemical shifts are believed to be due to the fact that in **175** (*meta*-linked) and **177** (*ortho*-linked) the bulky tosylamide groups sterically impinge on the outer *m*-terphenyl rings, which consequently relieves the strain by tilting toward each other (see Section 4.3), thus placing H$_{2'}$ in their combined shielding zones. This effect is not observed in **176** because the *p*-orientation of the linking units holds the outer *m*-terphenyl rings apart.

The cupped arrangement of the rings in **176** is evident from the ^{1}H NMR spectrum, which in all respects is similar to that of the cupped analogues **111** and **150**. The cupped arrangement of the rings in **175** is apparent from the chemical shift of the isolated proton on the xylylene ring, which appeared shielded at $\delta 6.49$; this shielding is due to the diamagnetic ring current of the central *m*-terphenyl ring.

The ^{1}H NMR spectrum of the flattened *ortho*-xylylene linked macrocycle **177** at RT showed broad featureless signals for most of the aromatic and methylene protons, indicating conformational flexibility at RT. (Neither **175** nor **176** showed any significant change in NMR spectrum down to $-70\,°C$.) However, the signals due to the protons on the central m-terphenyl ring ($H_{2'}, H_{4'}, H_{5'}, H_{6'}$) could be uniquely assigned. $H_{2'}$, as already mentioned, is shielded by the two outer m-terphenyl rings, which are tilted toward each other. Such a tilt is essential to accommodate the tosyl rings, which occupy space below the surface of the outer m-terphenyl rings and consequently influence the chemical shift of $H_{4'}$ and $H_{6'}$ at RT ($\delta\,5.72$). The magnetic equivalence of $H_{4'}$ and $H_{6'}$ and the broad signals due to the diastereotopic bridge methylene protons indicate a conformational process in which diagonally opposite bridges have identical spatial dispositions, and the conformational process may be represented by the interconversion of **177A** and **177B** (Scheme 22). In readily equilibrating (at RT) conformers **177A** and **177B**, the diagonally opposite bridges are either folded out or folded in. While the tosyl groups on the 'folded out' bridges sterically impinge on the outer m-terphenyl rings causing steric deshielding of the relevant protons (shown on the structure), the tosyl groups of the 'folded in' appendages lie below the surface of the outer m-terphenyl rings and thus diamagnetically shield the protons on the central m-terphenyl ring. On heating a sample of **177** to 54 °C, lines for all the resonances sharpened as a consequence of rapid interconversion of **177A** and **177B**. By cooling a solution of **177** to $-56\,°C$ it was possible to observe the ^1H NMR spectrum of a frozen single conformer. At that temperature, there were two sets of equal intensity signals for the tosyl ring protons as well as the methyl protons. The methylene resonances, however, being relatively sharp were not fully resolved with the expected eight sets of multiplets. Significantly, the $H_{4'}, H_{6'}$ signal remained symmetric, conforming to the earlier assumption that the diagonally opposite bridges have the same spatial disposition. The free energy ($\Delta H^{\ddagger}_c$) for the interconversion of **177A** and **177B** was determined to be 12.3 kcal/mol following the coalesence of the tosyl methyl singlets ($T_c = 246K$).

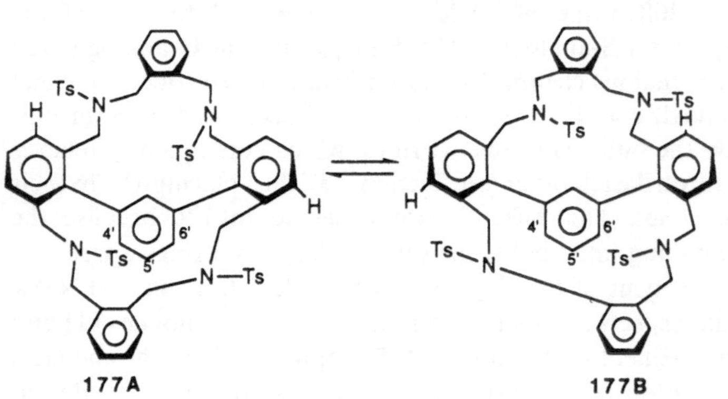

Scheme 22

The cupped tetrakis-N-tosylamides **175** and **176** were also prepared by coupling the tetrakis-N-tosylamide template **93** and the appropriate xylylene

$$(18)$$

dihalide [Eq. (18)] [11]. Neither a decrease nor an improvement of the yields of the coupled products was observed as a result of the switch in the electrophile-nucleophile role of these coupling partners. The advantage of having 8 (Scheme 21) as well as 93 [Eq. (18)] as base templates for the construction of cupped tetraazophanes is the fact that one now has a choice between using a dihalo or diamino derivative as the wall component in a coupling reaction.

Detosylation of 175, 176, and 177 provided the tetraamines 178, 179 and 180 respectively. The detosylation procedure developed by Trost [35] using Na-Hg in a buffered medium was effective for 175–177 (Scheme 21) and gave the corresponding tetraamines 178–180 (structures 13) in good to excellent yields. Another successful detosylation procedure was the 'excess LAH' [36] (50–60 fold excess) route, which is usually carried out under reflux conditions.

Structures 13

The structural (diagnostic NMR) features of 178–180 were quite similar to their corresponding S- and O-analogues. The appearance of the internal proton in 178 ($\delta 6.48$) and 180($\approx \delta 7.1$) in the expected ranges is consistent with the proposal (vide supra) that the bulky tosyl groups caused the abnormal shifts in their respective precursors 175 and 177.

4.6 Hydrocarbon Cuppedophanes via Sulfone Pyrolysis

Pyrolysis of poly-sulfones at elevated temperatures under reduced pressure is an efficient and proven method for the synthesis of hydrocarbon cyclophanes with

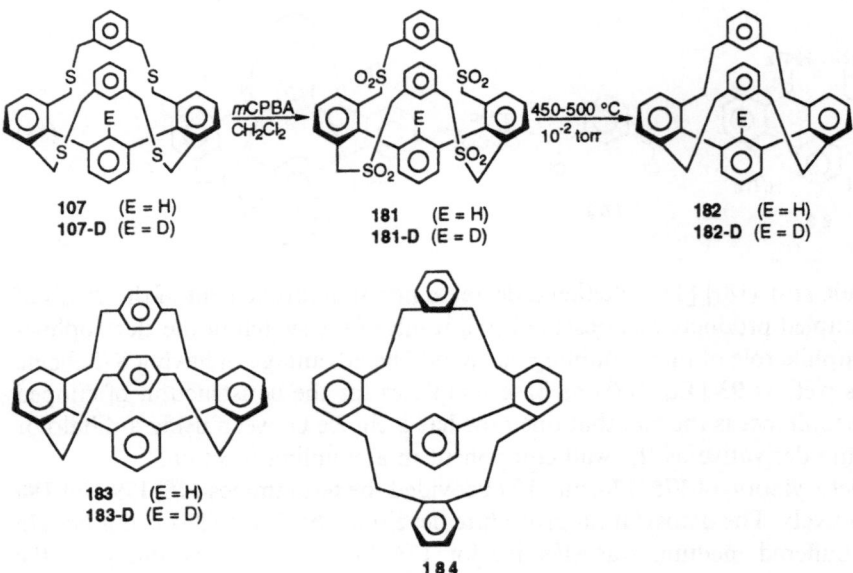

Scheme 23

multiple bridges [37]. The sulfone pyrolysis approach (involving a four-fold SO_2 elimination) was therefore used for the synthesis of a few representative members of hydrocarbon cuppedophanes. Oxidation of **107** with m-chloroperbenzoic acid (m-CPBA) in HOAc-CH_2Cl_2 at RT gave a nearly quantitative yield of the tetrasulfone **181** which on pyrolysis (450–500 °C) under reduced pressure gave a 27% yield of the hydrocarbon cuppedophane **182** (Scheme 23). An X-ray structure (Fig. 4) of **182** verified the structure and showed that the m-xylylene

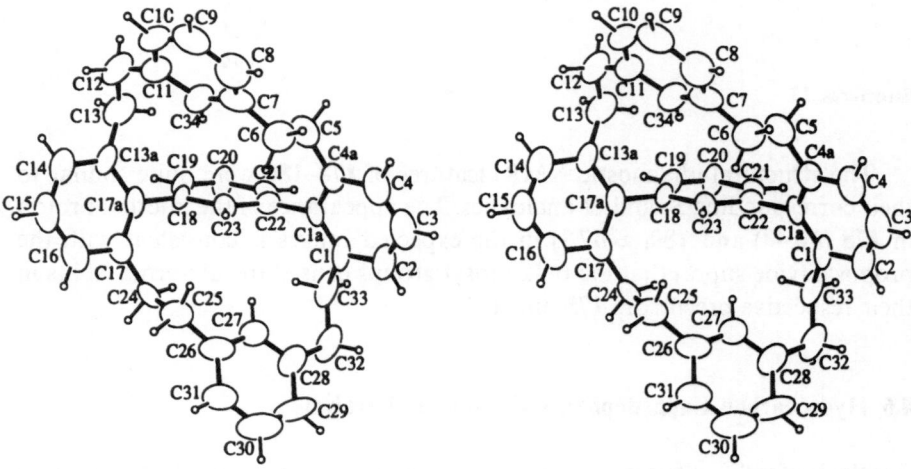

Fig. 4. Stereoview of cupped hydrocarbon **182**

rings on the tethers are in a face-to-face conformation with a slight lateral shift. The structure also rationalizes the substantial shielding of the internal hydrogen ($H_{2'}$), which appeared at $\delta 5.70$. This peak is absent in the spectrum of **182-D**, confirming the assignment.

Successful four-fold sulfone eliminations also led to the *p*-xylylene analogue **183** and the flattened *o*-xylylene bridged hydrocarbon **184** (Scheme 23). Hydrocarbon **183** has a cupped arrangement of rings, evident from the appearance of the $H_{2'}$ resonance at $\delta 6.59$, a signal that is absent in the spectrum of **183-D**. The *o*-xylylene–linked hydrocarbon **184**, as expected, has no unique aromatic resonances.

5 Salient NMR Features of Cuppedophanes and Their Significance in Conformational Assignments

The cuppedophanes discussed in the preceding sections can be classified into three different general structures, **A**, **B** and **C** shown in Scheme 24. The oval-shaped component in the bridges refers to the arene component present in the tether. These arene components, except in the case of the *o*-xylylene unit, carry a diagnostic proton (H_a) which can be influenced, depending on its position by the

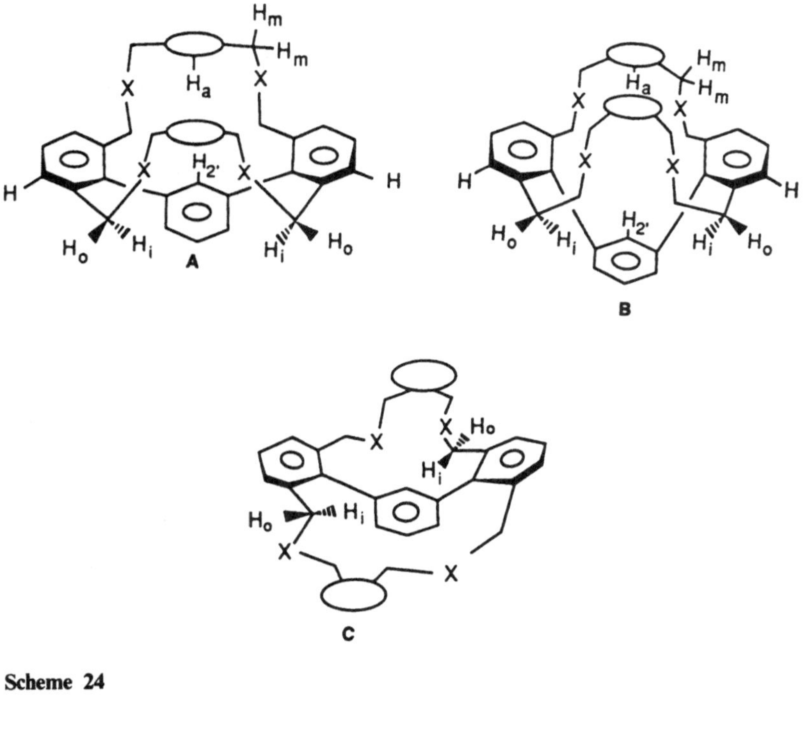

Scheme 24

155

diamagnetic ring current of the central *m*-terphenyl ring. The methylene unit attached to the *m*-terphenyl base bears a pair of diastereotopic protons of which H_i, the proton pointing inward, can also be influenced by the magnetic ring current of the central *m*-terphenyl ring. H_o, the proton pointing outward, lies almost in the same plane as the protons (shown on the structures **A**, **B** and **C**) on the outer *m*-terphenyl ring and consequently is not influenced by the ring current of any of the *m*-terphenyl rings. If 'X' bears large substituents (for e.g. the *p*-toluenesulfonyl groups in **175**, **176** and **177**) which can sterically interact with the outer *m*-terphenyl rings then a distortion of the cupped structure occurs with a tilt of the outer *m*-terphenyl rings towards each other (structure **B**).

The consequences of such a tilt are two-fold. The internal hydrogen ($H_{2'}$) is brought well within the combined shielding zones of the outer *m*-terphenyl rings

Table 2. Chemical shift values of significant protons invoked in conformational assignments of cuppedophanes

Structure (No, Type)	$\delta H_{2'}$	δHa	$\delta H_i + H_o$ (m, J in Hz)	δH_w (m, J in Hz)
(107, A)	6.39	6.74	3.36, 3.04 (ABq, J = 11.1)	3.63, 8.51 (ABq, J = 14.1)
(111, A)	6.26	6.83[a]	3.04, 2.39 (ABq, J = 10.8)	3.73, 3.63 (ABq, J = 13.4)
(112, C)	7.1–7.50[b]	–	3.63 (s)	3.77, 3.49 (ABq, J = 11.8)
(115, A)	6.35	6.33	3.37, 2.96 (ABq, J = 11.0)	3.54, 3.48 (ABq, J = 14.3)

Table 2. (Contd.)

Structure (**No, Type**)	$\delta H_{2'}$	δHa	$\delta H_i + H_o$ (m, J in Hz)	δH_w (m, J in Hz)
(**116, A**)	6.37	6.29	3.40, 2.96 (ABq, $J = 10.6$)	3.49, 3.43 (ABq, $J = 14.1$)
(**119, A**)	6.54	–	3.58, 3.09 (2d, $J = 10.4$)	3.72, 3.67 (ABq, $J = 13.9$)
(**118, B**)	5.03	–	2.85, 1.61 (2d, $J = 10.9$)	5.51, 4.42 (2d, $J = 13.7$)
(**120, A**)	7.4	–	3.36, 3.10 (2d, $J = 15.4$)	3.61, 3.51 (2d, $J = 12.3$ Hz)
(**185, A**)	–	6.71	3.47, 2.87 (2d, $J = 10.8$)	3.66, 3.46 (ABq, $J = 14.4$)
(**186, A**)	–	6.21	2.96, 2.70 (2d, $J = 10.5$)	3.49, 3.37 (ABq, $J = 14.4$)

Table 2. (Contd.)

Structure (No, Type)	$\delta H_{2'}$	δHa	$\delta H_i + H_o$ (m, J in Hz)	δH_w (m, J in Hz)
(124, A)	–	6.64	3.44, 2.83 (2d, J = 10.5)	3.64, 3.45 (2d, J = 14.0)
(125, A)	–	6.65	3.46, 2.86 (2d, J = 10.5)	3.59, 3.45 (2d, J = 14.0)
(127, A)	–	6.97	3.11, 2.28 (2d, J = 10.8)	3.75, 3.67 (2d, J = 14.0)
(126, A)	–	6.89[a]	3.18, 2.38 (2d, J = 9.9)	3.80, 3.67 (2d, J = 13.0)
(128, B)	–	–	3.03, 1.43 (2d, J = 9.0)	5.12, 4.37 (2d, J = 13.5)
(129, B)	–	–	3.06, 1.41 (2d, J = 10.0)	5.12, 4.37 (2d, J = 14.0)

Table 2. (Contd.)

Structure (No, Type)	$\delta H_{2'}$	δHa	$\delta H_i + H_o$ (m, J in Hz)	δH_w (m, J in Hz)
(131, B)	–	–	3.63, 3.83 (2d, J = 14.0)	3.81 (s)
(130, B)	–	–	3.68 (d, J = 15.0)	3.90, 3.682 2d, J = 16.0)
(133)	–	–	3.46, 3.70 (2d, J = 12.0)	3.56, 3.85 (2d, J = 14.0)
(143, B)	6.60	7.43	–	5.02, 5.09 (ABq, J = 12.3)
(145, B)	3.84	–	–	4.77, 5.46 (ABq, J = 11.5)
(146)	5.70	–	–	4.70, 5.48 (2d, J = 11.5) 4.73, 5.10 (2d, J = 11.5)

Table 2. (Contd.)

Structure (No, Type)	δH$_{2'}$	δHa	δH$_i$ + H$_o$ (m, J in Hz)	δH$_w$ (m, J in Hz)
(149, A)	6.71	6.95	3.98 (s)	4.50, 4.59 (ABq, $J = 12.5$)
(150, A)	6.21	7.06[a]	3.33, 3.46 (ABq, $J = 11.0$)	4.60, 4.62 (ABq, $J = 11.7$)
(151, C)	7.19	–	4.33, 4.30 (ABq, $J = 10.7$)	4.40, 4.44 (ABq, $J = 10.7$)
(175, B)	4.56	6.49	3.23, 4.44 (ABq, $J = 17.3$)	4.04, 4.42 (ABq, $J = 14.4$)
(176)	6.25	6.72[a]	2.75, 4.25 (2d, $J = 18.4$)	3.92, 4.98 (ABq, $J = 14.0$)

Table 2. (Contd.)

Structure (**No, Type**)	$\delta H_{2'}$	δHa	$\delta H_i + H_o$ (m, J in Hz)	δH_w (m, J in Hz)
(**177, C**)	4.76[c]	–	3.33, 4.19[c] (ABq, $J = 15.4$)	3.44, 3.91[c] (ABq, $J = 15.4$)
(**178, A**)	6.48	6.75	3.26 (s)	3.77(s)
(**179, A**)	6.27	6.93[a]	2.51, 2.71 (2d, $J = 12.1$)	3.62, 3.97 (2d, $J = 13.4$)
(**180, C**)	7.04–7.50[b]	–	3.47, 3.65 (ABq, $J = 11.3$)	3.64, 3.76 (ABq, $J = 11.9$)

[a] Chemical shift of the 'bottom' protons; see Sect. 4.1.
[b] Chemical shift is part of the multiplet.
[c] Data from spectrum measured at 54 °C.

and the wall components of the structure are raised affecting the chemical shifts of both H_a and H_i. The tilt of the outer rings toward each other also occurs when the wall components are anchored directly to the outer rings as in tetraoxaphanes **143** [Eq. (14)], **145** and **146** (Scheme 15) or when the tether carries multiring arenes as in **118** [Eq. (10)], **128**, **129**, etc. In these phanes the unfavorable interaction that is relieved by inward bending of the outer rings is the steric contact between the rings on the tether and the central *m*-terphenyl ring. The flattened structure c is found exclusively where the arene ring on the tethers is an *o*-xylylene unit. Even in such flattened structures, the above-mentioned tilt of the outer *m*-terphenyl rings occurs if X bears a bulky substituent (for e.g. 177). Table 2 lists the chemical shifts of significant protons, $H_{2'}$, H_a, $H_i + H_o$ and H_w, the diastereotopic methylene protons nearest to the wall component.

6 Cappedophanes

6.1 Synthesis and Reactions of Tetrathiacappedophanes

Cappedophanes (general structure 2) (Structures 1) are a unique class of molecules with the $C_{2'}$ substituent of the *m*-terphenyl base encapsulated. These molecules are therefore potential candidates to study functional group properties within a microenvironment. The first two members that were prepared are the tetrathiacappedophanes 188 and 189 [9] (Scheme 25). Coupling of the tetrakis (bromomethyl)-*m*-terphenyl template 8 with tetrathiol 187 gave cappedophanes 188 and 189 in 10% and 1.9% yields respectively.

The low yields of these coupled products is a reflection of the difficulty in bridging across each external *m*-terphenyl ring. The $C_{2'}$-H ($H_{2'}$) of both 188 and 189 is placed in close proximity to the capping arene ring (estimated to be within 2.16 Å from the mean plane of the cap) and is therefore influenced by its diamagnetic ring current. The upfield chemical shifts of $H_{2'}$ in 188 and 189 ($\delta 3.97$ and $\delta 4.23$; these signals were absent in the deutero analogues 188-D and 189-D) is a consequence of shielding contributions from the capping ring as well as from the tilted outer rings of the *m*-terphenyl base. The extent of these individual contributions is hard to predict but it is suggested (by additional

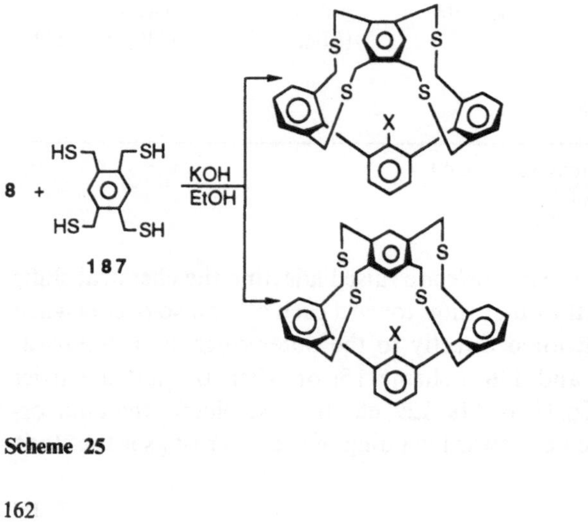

188	(X = H)
188-D	(X = D)

189	(X = H)
189-D	(X = D)

Scheme 25

examples, vide infra) that the tilted outer *m*-terphenyl rings contribute at least as much as the capping ring toward the shielding of these protons.

Cappedophanes **188** and **189** can be easily distinguished and characterized by the chemical shift of the isolated proton on the capping ring. In **188** the proton is juxtaposed on top of the tilted outer *m*-terphenyl rings and appears shielded at $\delta 4.75$. The extent of the tilt of the outer *m*-terphenyl ring can be gauged from the fact that an identical proton in the analogous structure **190** [38] (Structure 14) appears shielded at $\delta 5.03$ even though the distance between this proton and the parallel rings responsible for shielding is considerably shorter due to the absence of the sulfur atom on the bridges. The isolated protons on the capping ring of **189** point away from the outer *m*-terphenyl rings and appear at $\delta 8.30$, slightly deshielded in this case.

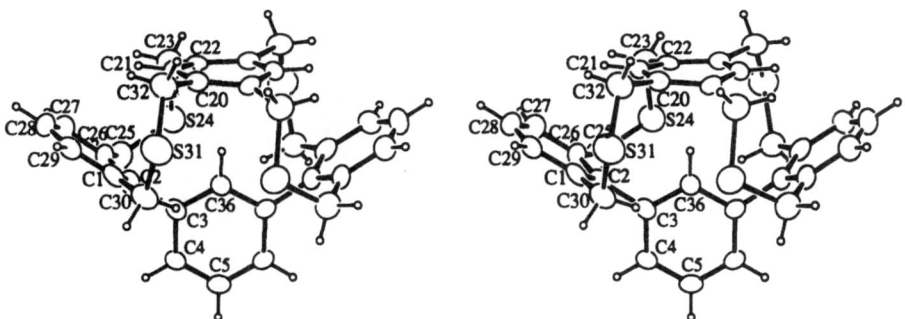

190

Structure 14

An X-ray structure determination on a single crystal of **188** verified the cappedophane structure (Fig. 5) and exhibited some interesting features. Although one might have expected **188** to have C_{2v} symmetry from its structural formula, distortion of the capping ring and the connecting arms reduced the observed symmetry to C_2. The capping unit is not planar and adopts a pseudoboat conformation. Two opposite methylene carbons attached to this ring (C_{32}, at the front left and rear right in Fig. 5) lie above its mean plane, and the other two methylene carbons (C_{23}, front right and left rear in Fig. 5) lie below that plane. The two $-CH_2SCH_2-$dihedral angles in the two types of arms

Fig. 5. Stereoview of **188** showing nonplanarity of the cap and twisting of the linking arms

are consequently different from each other. Further examination of the X-ray structure reveals that the volume enclosed between the *m*-terphenyl base and the cap is too small to accommodate C_2'-substituents other than hydrogen. The vaulted phanes (Sec. 6.3) are specialized cases of cappedophanes with a larger enclosed volume between the cap and the base.

The tetrathiacappedophane **188** was oxidized to tetrasulfone **191** using *m*-CPBA and the latter was pyrolysed to yield the hydrocarbon cappedophane **192** and the disulfone **193** [9] (Scheme 26). The hydrocarbon **192**, with all four of its tethers contracted by the removal of sulfurs, serves as a good model to study distortions of arene rings, a major theme of cyclophane research. The chemical shift of the diagnostic proton H_2' of **192** appeared at $\delta 3.77$, i.e. an enhanced shielding of about 0.2 ppm from that of the corresponding proton in **188** (Scheme 25). This is in contrast to an increased shielding of 1.08 ppm experienced by the isolated proton on the cap, which appeared at $\delta 3.67$. This discrepancy clearly points to a distorted capping ring which sustains only minimal ring current.

A solid state X-ray structure of **192** has not been made to confirm the distorted nature of the capping ring. However the assumption is further validated by the NMR features of the less strained disulfone **193**. The internal hydrogen H_2' of **193** appeared at $\delta 3.31$, the most shielded in this series of cappedophanes. In this particular case H_2' experiences a shielding contribution from an outer-*m*-terphenyl ring as well as the capping ring. Table 3 lists the

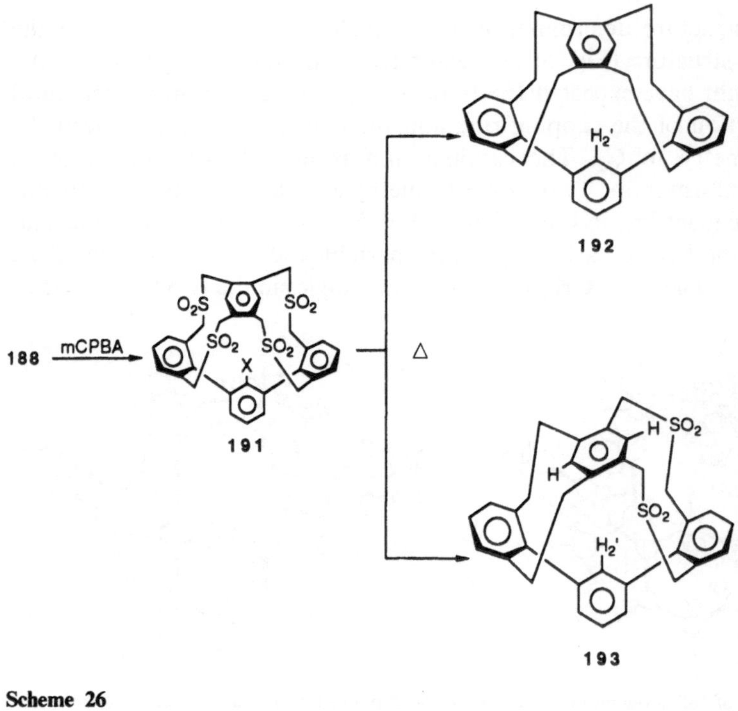

Scheme 26

Table 3. Chemical shift values of diagnostic protons in cappedophanes

Compound	$\delta H_{2'}$	δH (isolated proton on the capping ring)
188	3.97	4.75
189	4.23	8.30
192	3.77	3.67
193	3.31	3.45 (between hydrocarbon bridges)
		5.11 (between SO_2 bridges)

chemical shifts of the internal protons ($H_{2'}$) and the isolated proton on the capping ring of various cappedophanes.

Close examination of the table reveals two factors. (i) The outer *m*-terphenyl rings significantly contribute toward the shielding of both the internal hydrogen ($H_{2'}$) and the isolated capping protons, and (ii) the extent of the shielding contribution from the capping ring decreases as the bridge contractions occur.

6.2 Attempted Synthesis of Cappedophanes with Alkenyl Caps and an Unusual Rearrangement of α,α'-Dithiacuppedophane Ketones

Cappedophanes with alkenyl caps, as in general structures **194** and **195** (Structures 15), are interesting molecules for the following reasons: (i) a comparison of the chemical shifts of the diagnostic proton ($H_{2'}$) in **194** or **195** with the corresponding signals in cappedophanes with arene caps (Section 6.1) will help to delineate a comparison between ring current effects of arene rings and those of the alkenyl double bonds, and (ii) the alkenyl cappedophanes **194** and **195** can be precursors to other theoretically interesting molecules with either a cyclopropyl or an epoxide cap. Direct approaches to **194** and **195** by coupling the

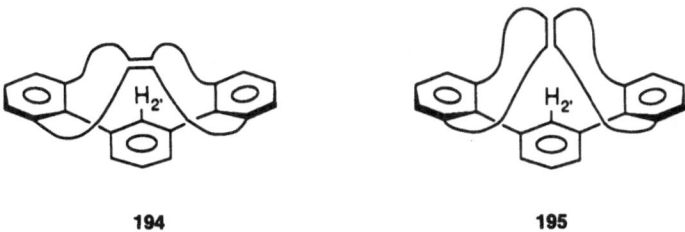

194   **195**

Structures 15

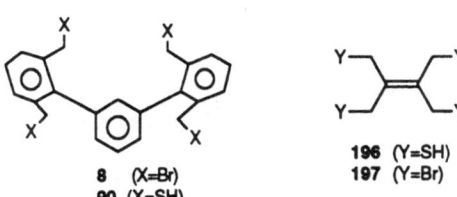

8 (X=Br)
90 (X=SH)

196 (Y=SH)
197 (Y=Br)

Structures 16

tetrakis(bromomethyl)-m-terphenyl 8 with tetrathiol 196, or the tetrakis(mercaptomethyl)-m-terphenyl 90 with tetrabromide 197 (Structures 16) failed to give the desired cappedophanes [26]. An alternate approach involving a McMurry coupling of the diketone 120 was sought and it was during the synthesis of this diketone that the unexpected rearrangement of α,α'-dithiaketones was discovered [26].

Treatment of tetrathiol 90 (Structures 16) with 1,3-dichloroacetone (198) in a suspension of cesium carbonate in DMF gave a 65% yield of the desired diketone 120 (Scheme 27). The structure of 120 was evident from its NMR spectrum and from a solid state structure determination carried out on a single crystal (vide infra).

An alternate synthesis of 120 employing NaOMe in MeOH (in place of Cs_2CO_3 in DMF) as the base for coupling 90 with 1,3-dichloroacetone gave two new products 199 and 200 [26] in addition to the desired diketone 120 (Scheme 27). The structures of 199 and 200 were deduced from their respective NMR spectra and in the case of 200 an X-ray structure was also carried out (vide infra). Separate treatment of 120 as well as 199 with NaOMe in MeOH gave the bisthioketal 200, proving that in the coupling reaction 120, 199 and 200 are formed sequentially.

A mechanistic pathway by which 120 eliminates the elements of ketene to provide 199 and 200 is intriguing and was investigated by looking at the fate of several model compounds under similar reaction conditions. The m-terphenyl dithiaketone 201, a monocyclic analogue of 120, as well as the tetrathia diketones 202 and 203 (Structures 17) were inert to rearrangements under reaction conditions identical to those in which 120 underwent molecular fragmentation. This suggested that the rearrangement observed with 120 is not

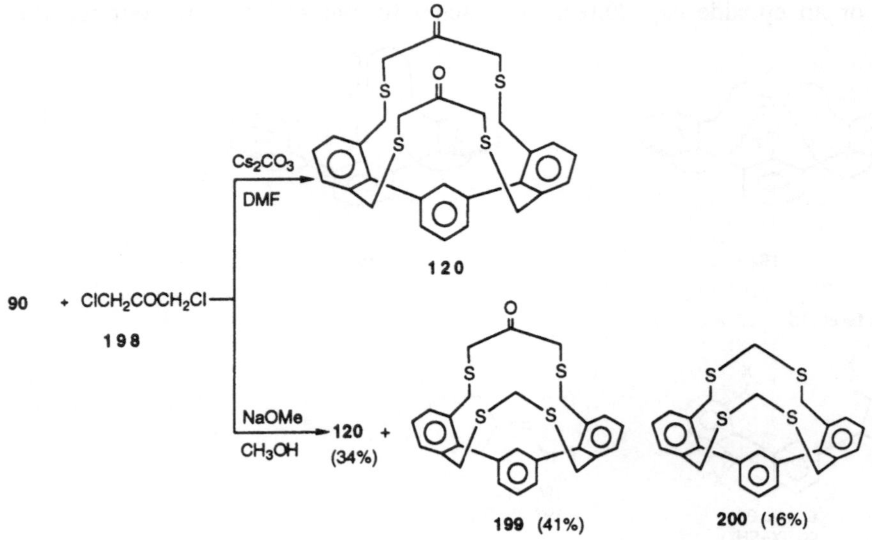

Scheme 27

Structures 17

general for α,α'-dithiaketones. However, during the synthesis of the model compound **204** from 1,2-bis(mercaptomethyl)benzene **110** and 1,3-dichloroacetone using NaOMe in MeOH provided, in addition to **204**, the dithiaketone **205** and the acetyl thioketal **206** (Structures 17). The formation of **206** in this coupling reaction is indicative of a rearrangement process similar to that which occurred with **120** (Scheme 28). The fact that model compounds **201, 202, 203** and **204** do not exhibit any propensity for the rearrangement and the acetyl thioketal **206** does not eliminate ketene upon further treatment with NaOMe in MeOH indicates that the observed rearrangement is specific to the doubly bridged *m*-terphenyl systems.

The mechanism shown in Scheme 28 is only speculative at the moment. Removal of an α protons by base should give the anion **A**. The bicyclic ketone **120** has two sets of α-protons and it is not known whether the proton removal is stereoselective. Anion **A** might then displace directly on sulfur to give **B** (path *a*) or arrive at **B** via a Favoroskii-type intermediate **C** (path *b*), although there seems to be no special advantage to the latter route. The intermediate **B** might eliminate ketene to give the anion of **199** directly or more likely accept a proton to give **D** which might then be cleaved by attack at the carbonyl group. The solid state structure of **120** (Fig. 6) was carried out to scrutinize the pertinent dihedral angles of –S–CH$_2$–C(O)–CH$_2$– linkages because they could determine whether a carbanion formed on one side of the carbonyl group could approach the sulfur atom on the other side of the carbonyl group.

The solid state structure of **120** (Fig. 6) shows that one of the two dithiaketone arms (from S6 to C11) that connect the outer rings of the *m*-terphenyl moiety is disordered. The structure was solved and refined (R = 0.041) with an occupancy

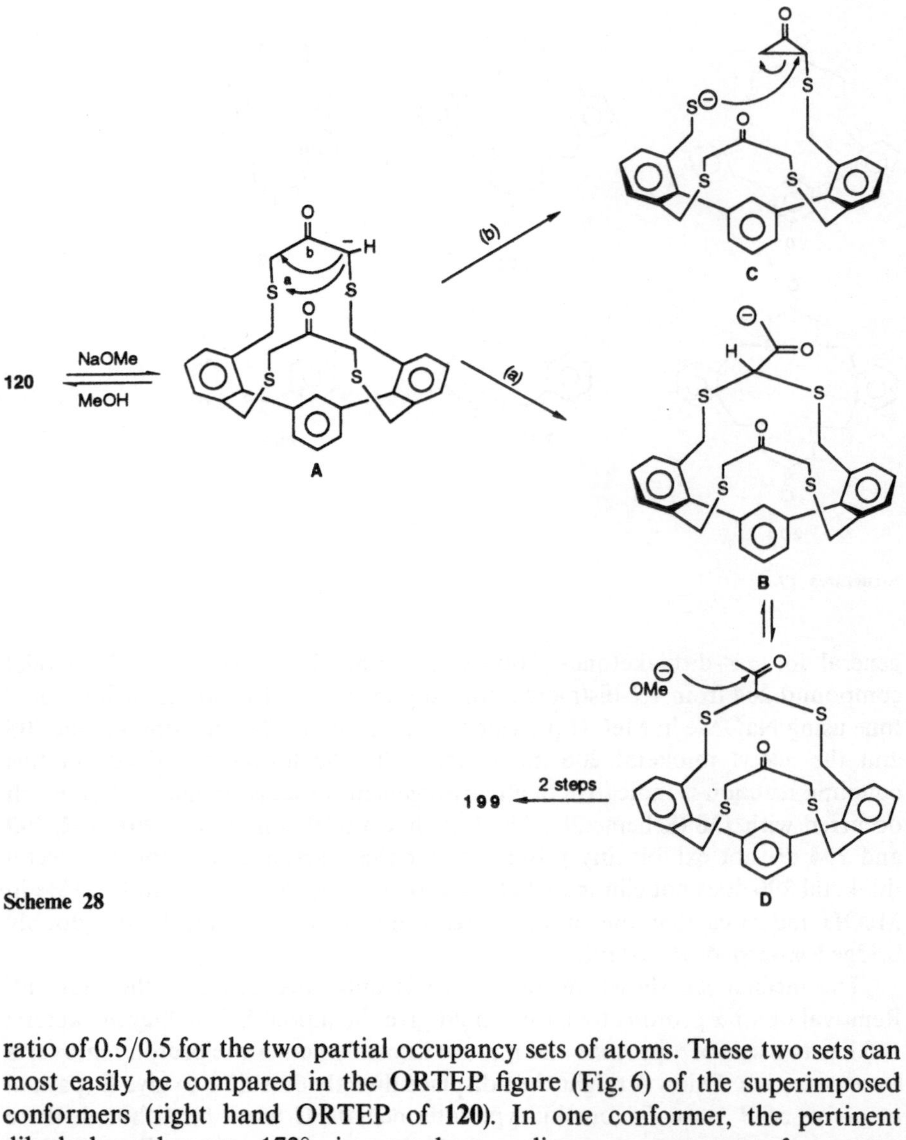

Scheme 28

ratio of 0.5/0.5 for the two partial occupancy sets of atoms. These two sets can most easily be compared in the ORTEP figure (Fig. 6) of the superimposed conformers (right hand ORTEP of **120**). In one conformer, the pertinent dihedral angles are 172°, i.e. nearly a colinear arrangement for atoms $-S-CH_2-C(O)-CH_2$. In this conformer the approach of either C9' to S6 or C7' to S10 is impossible. However in the other conformer one of these dihedral angles is 90° ($S_{10}-C_9-C_8-C_7$) and the other ($S_6-C_7-C_8-C_9$) is 173°. Also in the fixed arm, one of these angles ($S_{23}-C_{24}-C_{25}-C_{26}$) is 85° and the other ($S_{27}-C_{26}-C_{25}-C_{24}$) is 147°. Even though the dihedral angle of approximately 90° may allow some carbanion–sulfur 1,4-interactions, a smaller dihedral angle found between the same four atoms in model compounds **202** and **203** indicated that the occurrence of the rearrangement has other, yet to be understood, factors controlling it.

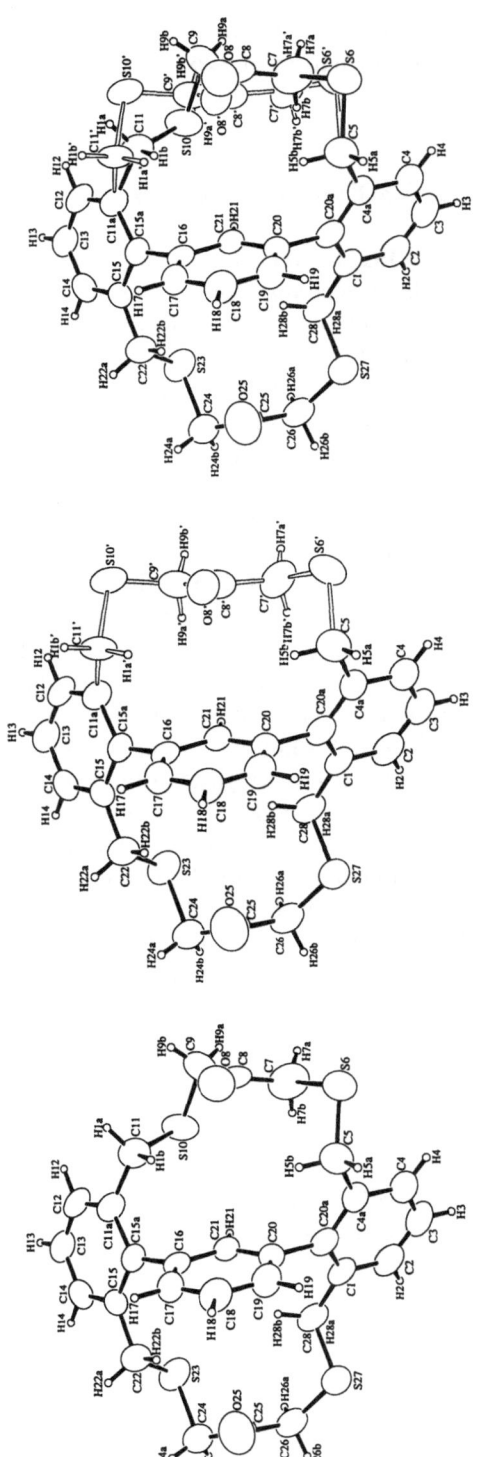

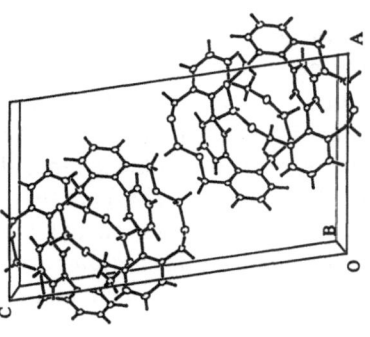

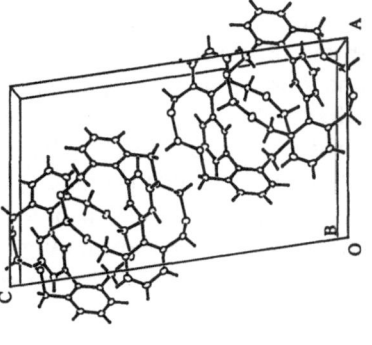

Fig. 6. Solid state structures of **120** (top, 3 views, see text) and **200** (bottom)

169

Although the synthesis of the bicyclic diketone **120** led to the discovery of novel rearrangements mentioned above, transformation of **120** to an alkenyl cappedophane has so far been unsuccessful.

6.3 Self-filled, Vaulted and Intracavity-Functionalized Cappedophanes

The cappedophanes **188, 189** (Scheme 25) and **192, 193** (Scheme 26) described in Sect. 6.1 enclose a cavity only large enough to accommodate a proton as a substituent on C_2. Enlarging the cavity to accommodate larger substituents on C_2 might be attained by raising the cap through increasing the length of the linkers, as shown in structure **207** (Structures 18). However CPK models of **207** with longer linkers between the cap and the molecular base showed that flexible links can easily attain collapsed conformations which diminish the cavity volume. Placing the cap on a rigid cuppedophane, as in structure **208** (Structures 18), avoids this possibility and leads to vaulted cappedophanes. The two most direct routes to vaulted phanes (hereafter designated with a **v** after the compound number, i.e. **208v**) are (a) to fabricate the walls and cap and then attach that unit to the m-terphenyl base in one step, or (b) include on the walls functionality suitable for later attaching the cap. The synthesis of vaulted phanes by coupling the m-terphenyl base **8** (Structures 16) with the tetrathiol vault **209** [17, 39] (a structural unit that combines the walls and a capping ring) showed a remarkable propensity for forming the self-filled conformer of the vaulted phane **208v**, namely **208sf** (sf denoting self-filled) (Scheme 29). m-Xylylene dithiols react with tetrakis(bromomethyl)-m-terphenyl **8** across the outer bridges to produce cuppedophanes (Sect. 4.1). Consequently it was expected that the linked tetrathiol **209** would react with **8**, via a four-centered high dilution coupling, to give the vaulted cappedophane **208v**. Even though the vaulted phane **208v** was formed (yield 2%), the major product (yield 62%) was the self-filled conformer **208sf**. Conformer **208sf** arises out of a displacement of the four bromine atoms of **8** from *below*, thus encapsulating the central m-terphenyl ring in the cavity; vaulted cappedophane **208v** would arise by an analogous displacement from *above*.

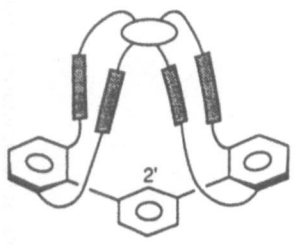

207 (▬▬ =several atoms)

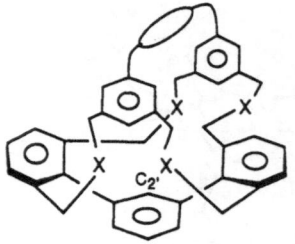

208

Structures 18

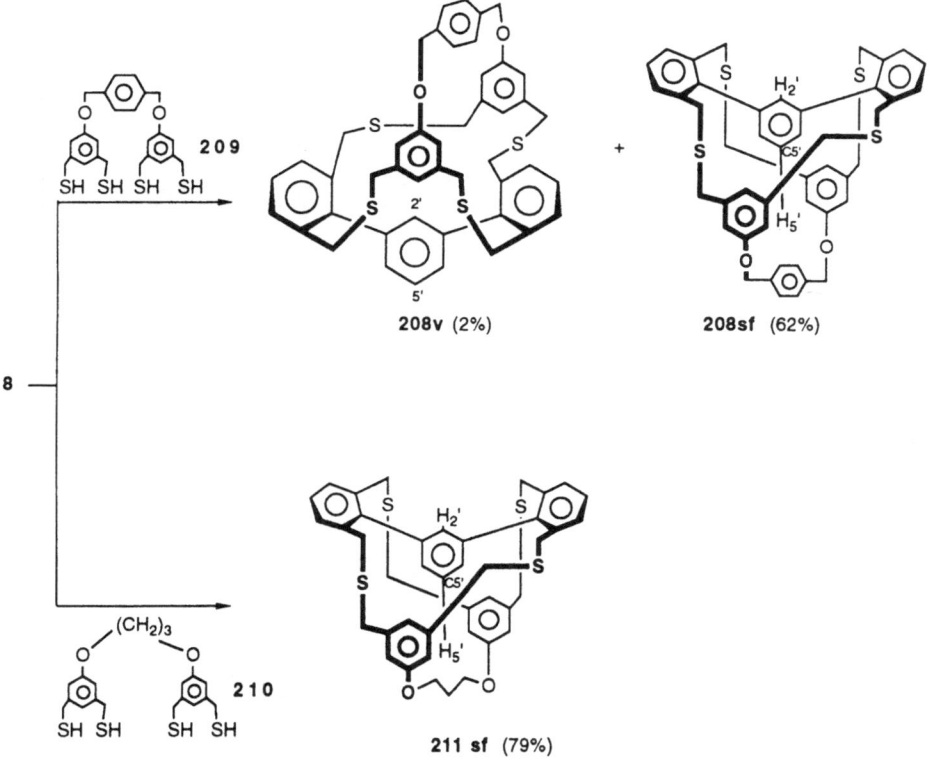

Scheme 29

The identification and characterization of these two conformers was accomplished by ¹H NMR and was clear as a consequence of unique resonances in the two spectra. In the self-filled conformer, proton $H_{5'}$ on the central *m*-terphenyl ring is located above the *p*-xylylene ring; consequently it is extremely shielded and appears at $\delta 4.31$ (t, $J = 7.7$ Hz). NOE experiments also indicated spatial closeness between $H_{5'}$ and the *p*-xylylene ring protons (10.4% enhancement for the *p*-xylylene ring protons). In contrast, $H_{5'}$ in **208v** appeared at $\delta 7.35$. The most diagnostic signal in the ¹H NMR spectrum of **208v** is a one-proton triplet ($J = 1.6$ Hz) at $\delta 5.70$ unique to $H_{2'}$. The upfield shift experienced by this proton is possibly due to a combination of two ring current effects, i.e. shielding contributions from the flanking 1,3,5-tri-substituted rings and from the *p*-xylylene ring at the top. In contrast, $H_{2'}$ in **208sf** appeared at the somewhat less shielded position of $\delta 6.24$.

The formation of **208sf** as the major coupling product of **8** and **209** was an unexpected observation and consequently reasons were sought for such a predominance. An energy minimization carried out on **208v** and **208sf** indicated that the self-filled conformer enjoys a van der Waals stabilization energy of 3.86 kcal/mol over the vaulted conformer. The van der Waals energy contributions toward the total energies of **208v** and **208sf** are 71.4 and 75.2 kcal/mol

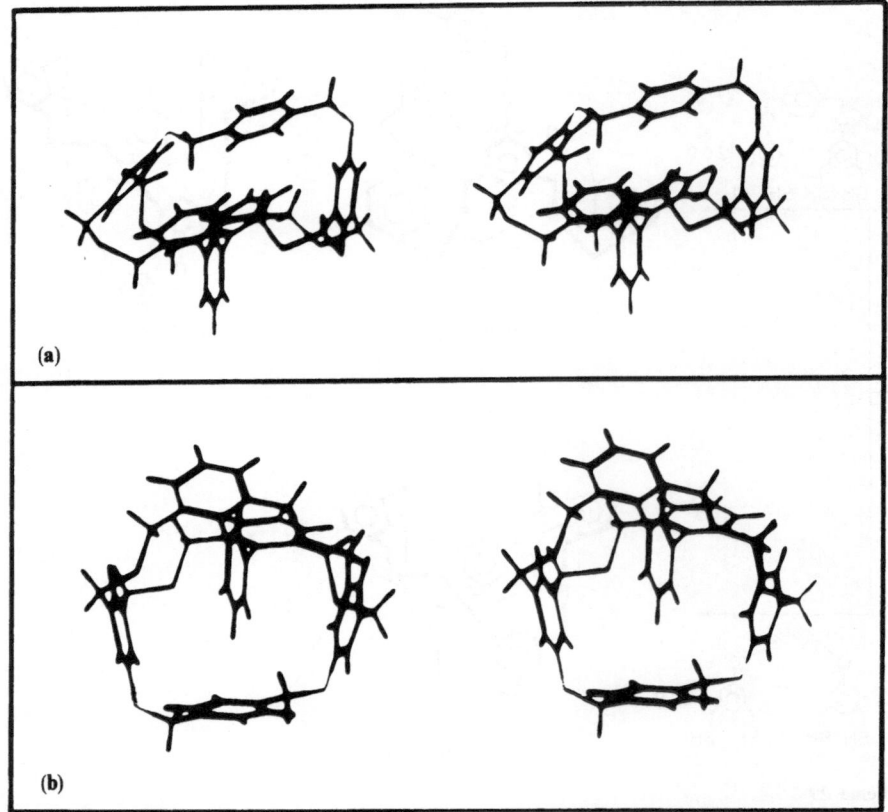

Fig. 7. Stereoviews of energy-minimized conformations of **208v** (a) and **208sf** (b)

respectively, and the difference favoring the self-filled conformation is similar to other reported values [40]. Stereoviews of the energy minimized conformations of **208v** and **208sf** are shown in Fig. 7.

Coupling of **8** with tetrathiol **210** provided only the self-encapsulated product **211sf** (Scheme 29) in spite of the shorter chain linking the two phenolic oxygens. The extent of the van der Waals stabilization of **211sf** was not calculated, but it seems likely that, due to the shorter chain linking the two phenolic rings, the effect is larger in this case, an assumption supported by the exclusive formation of **211sf**. The self-filled nature of **211sf** was evident from its ^{1}H NMR spectrum. The clearly distinguished $H_{5'}$ triplet ($J = 7.7$ Hz) at $\delta 6.23$, when irradiated resulted in a 10.8% intensity enhancement of the four-proton triplet ($\delta 4.37$) of the trimethylene unit. This observation, in addition to proving the close proximity of $H_{5'}$ to the trimethylene unit, also establishes the conformation of the trimethylene unit with its outer methylene units folded in (as drawn).

Examination of structure **208sf** (Fig. 7) suggested that by having bulky substituents on $C_{5'}$, the self-encapsulation could be prevented and the yields of

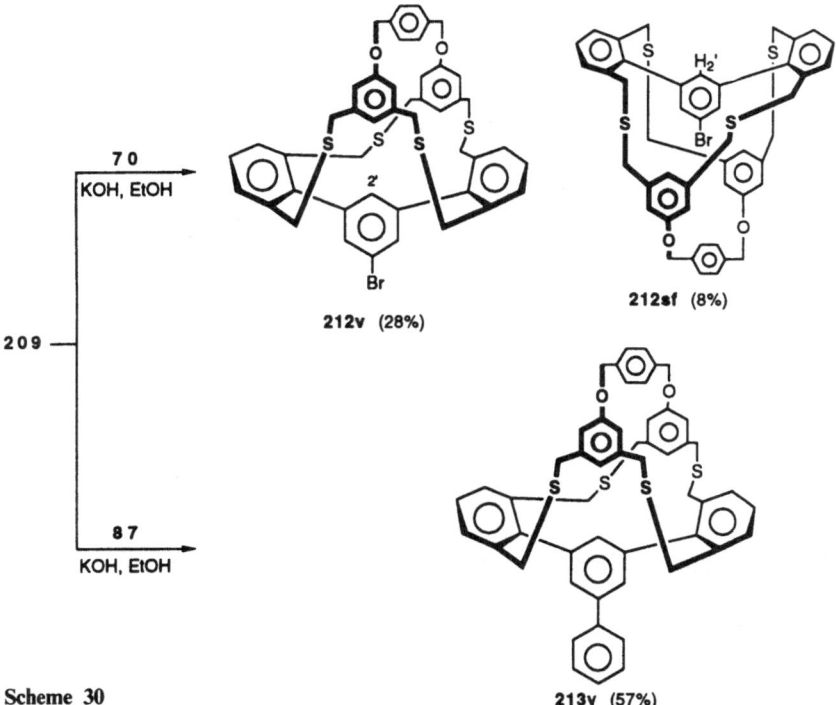

Scheme 30

vaulted phanes increased. *m*-Terphenyl based synthons **70** (Scheme 9) and **87** (Scheme 10), with bromine and phenyl substituents respectively at $C_{5'}$, were consequently coupled with the tetrathiol vault **209** (Scheme 29). Using **70** as the *m*-terphenyl base, both vaulted (**212v**) and self-filled (**212sf**) phanes were produced, with the vaulted conformer as the major product. The structural characterization of **212v** was made easy by the diagnostic chemical shift of $H_{2'}$ (δ 5.65); however the lack of a $C_{5'}$ proton made the unambiguous characterization of **212sf** difficult. The structure of **212sf** was proved chemically; treatment of **212sf** with *n*-BuLi at $-78\,^{\circ}$C followed by aqueous quench provided **208sf**. In addition to proving the structure, this experiment showed that the encapsulated functionality in **212sf** is accessible for chemical manipulations.

While the bulkiness of the $C_{5'}$-bromo substituent was insufficient to produce vaulted phanes exclusively, the phenyl substituent was large enough to exclude the formation of any self-filled conformer. The structure of **213v** (Scheme 30) was apparent from its nearly identical ^{1}H NMR spectrum to that of **208v**.

The utility of tetrathiol **209** in the construction of capped phanes is also illustrated by its coupling with tetrabromide **139** to give the remarkable caged phane **214** [19] [Eq. (19)].

The coupling reactions discussed so far in this Section involve a *m*-terphenyl base and a prefabricated vault, and revealed a propensity for the formation of self-filled over vaulted conformers. The alternate synthetic strategy, to attach a cap to a cuppedophane with wall components bearing suitable functionalities, was also attempted. Treatment of the diphenolic cuppedophane **116** with p-xylylene dibromide in DMF in the presence of K_2CO_3 did indeed provide a 55% yield of the vaulted phane **208v** [Eq. (20)]. However in spite of the predominance

of the vaulted phane in this coupling reaction, the intriguing presence of traces of the self-filled conformer was somewhat disturbing. Although thermal isomerization of the vaulted phane by a flip of the central m-terphenyl ring from outside the cavity to the inside could lead to self-filled phanes, it is unlikely that is what is occurring in this case. This conclusion is based on the results of two coupling reactions carried out at different temperatures (RT and $\approx 60\,°C$); each produced only trace amounts of the self-filled conformer **208sf**. Thermal isomerization of **208v** to **208sf** could in fact be observed, but occurred very slowly even at the temperature of refluxing DMF.

A more plausible explanation for the formation of **208sf** is that a conformational change occurs at some stage during the double alkylation of **116** (Scheme 31). Monoalkylated intermediate may cyclize to give **208v** or may undergo conformational changes (due to steric reasons) once or twice to give, after the second displacement, oligomers or the self-filled conformer.

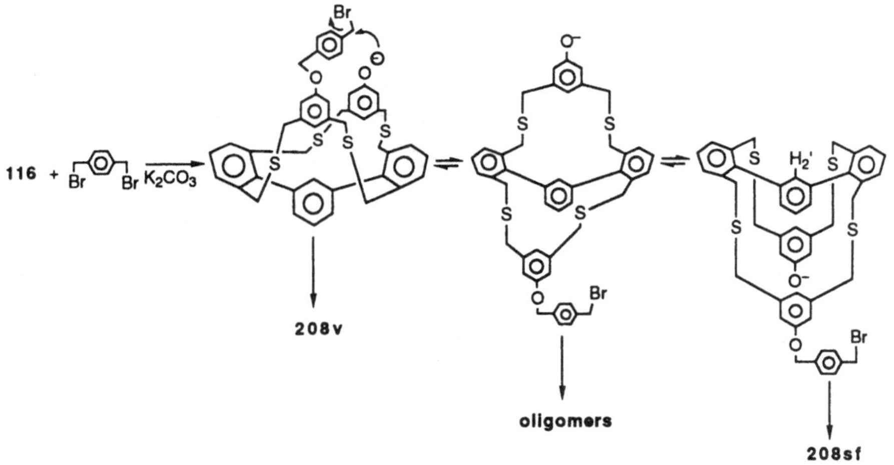

116 + [structure with Br, Br, K₂CO₃] →

208v

↓

oligomers

208sf

Scheme 31

The conformational flip is in part influenced by the presence of bulky internal (C₂·) substituents. A case in point is the exclusive formation of **215sf** from a coupling between the diphenolic cuppedophane **186** and *p*-xylylene dibromide [Eq. (21)]. The presence of a bromine atom at the internal position is believed to be responsible for the flip of the phenolic rings, leading to the sole formation of **215sf**. It has been previously observed in solid state structures of cuppedophanes [20] that a conformational flip of wall component rings occurs if the *m*-terphenyl unit bears internal bulky substituents (see Fig. 1).

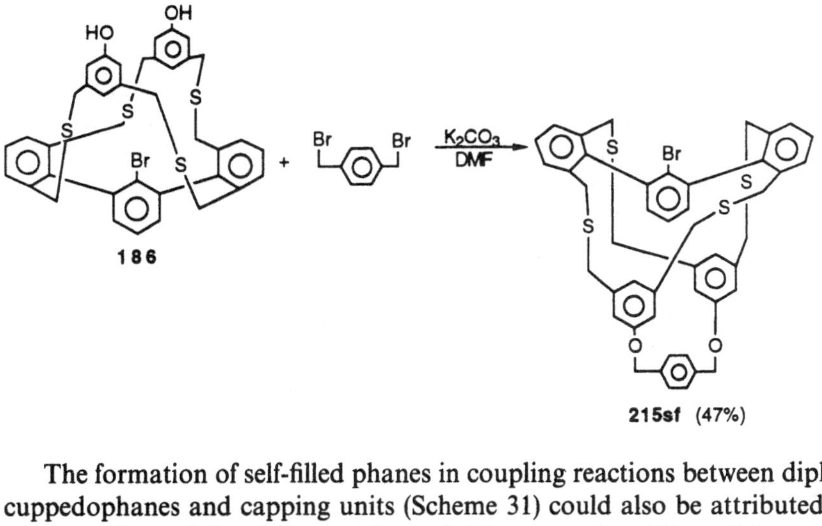

186 + [structure with Br, Br, K₂CO₃, DMF] → (21)

215sf (47%)

The formation of self-filled phanes in coupling reactions between diphenolic cuppedophanes and capping units (Scheme 31) could also be attributed to the nature and reactivity of the capping unit. For example, with *p*-xylylene dibromide as the capping unit, the rate of the second alkylation of the phenol was apparently faster than the rate of conformational flip, thus leading to the predominance of **208v**. However that was not the case when the capping unit

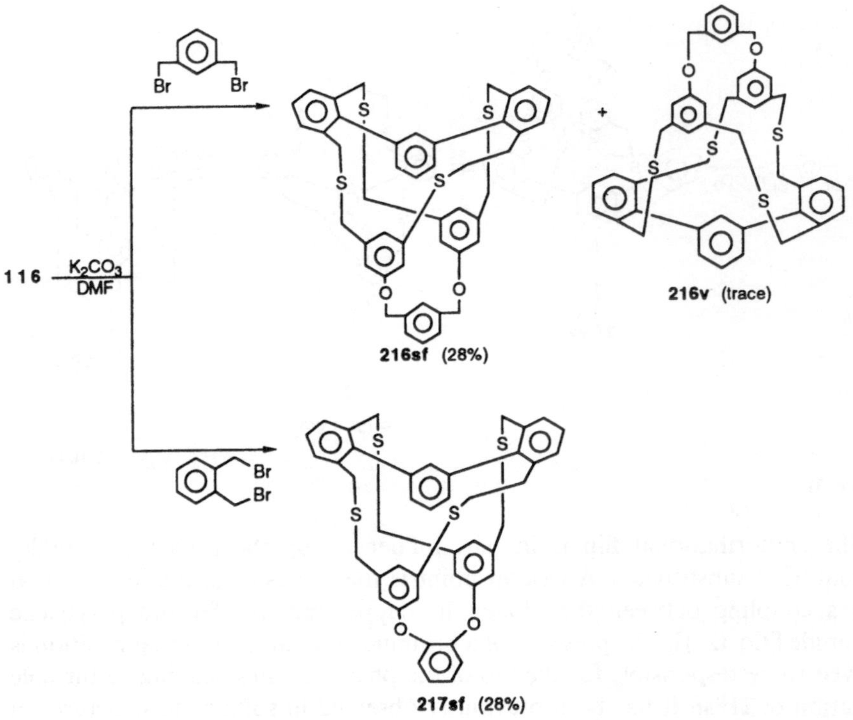

Scheme 32

was either *m*-xylylene dibromide or *o*-xylylene dibromide. With *m*-xylylene dibromide as the linker, the vaulted phane **216v** was only obtained in trace amounts, whereas the self-filled conformer **216sf** was produced in 28% yield (Scheme 32). The change in product ratio is presumably due to the shortening of the linker length, which causes the rate of the second alkylation (Scheme 31) to be slower than the conformational flip; thus **216sf** is the predominant product. The use of the shorter *o*-xylylene linker, as expected, provided only the self-filled phane **217sf** (28%) (Scheme 32). The structures of vaulted phane **216v** and the self-filled phanes **216sf** and **217sf** were characterized by the diamagnetic proton chemical shifts in their respective ^{1}H NMR spectra.

The results summarized in this section suggest that in order to construct vaulted cyclophanes with intracavity functionality it is necessary to have large groups (i.e. phenyl, as in **213v** Scheme 30) at $C_{5'}$ of the *m*-terphenyl unit.

7 Conclusions

In this review we have shown that the tandem-aryne synthesis (Sec. 3.1) provides a useful general entry to *m*-terphenyls suitably substituted for the construction of *cuppedophanes* and *cappedophanes*, as well as other macrocyclic phanes.

The *cuppedophanes* have cavities whose depth can easily be varied, and whose base and rim may contain functionality useful for binding. As yet, however, few binding studies have been carried out. It remains for the future to design cuppedophanes with specific geometries suitable for binding specific substrates. The syntheses are short, versatile, and the conformations of quite a few cuppedophanes are now well established, so that the groundwork for these future studies has been laid.

Cappedophanes are also quickly assembled using similar methods. The high propensity for these molecules to adopt conformations in which the cavity is self-filled has been established; so, too, are methods for avoiding such conformations. In this way, cappedophanes with functionality within the cavity can now be prepared, and the way is open to study the behavior of a given group (for example, the acidity of a carboxyl group) inside and outside a well-specified cavity.

The surface of cuppedo- and cappedophane chemistry has barely been scratched, and we hope that others will find this a fruitful area for research.

References

1. Tabushi I, Yamamura K, Kuroda, Y (1989) In: Atwood JL (ed) Inclusion phenomenon and molecular recognition. Plenum, New York, p 257
2. Gokel GW (1991) In: Atwood AL, Davies JED, MacNicol DD (eds) Inclusion compounds, vol 4. Oxford University Press, New York, p 283
3. Gutsche CD (1991) In: Atwood JL, Davies JED, MacNicol DD (eds) Inclusion compounds, vol 4. Oxford University Press, New York, p 27
4. Murukami Y (1989) In: Atwood JL (ed) Inclusion phenomenon and molecular recognition. Plenum, New York, p 107
5 Diederich F (1989) In: Atwood JL (ed) Inclusion phenomenon and molecular recognition. Plenum, New York, p 93
6. a. Du C-JF, Hart H, Ng K-KD (1986) J Org Chem 51: 3162; b. Du C-JF, Hart H. (1987) J Org Chem 52: 4311; c. Hart H, Ghosh T (1988) Tetrahedron Lett 29: 881
7. Cram DJ (1987) Science (Washington, DC) 219: 1177
8. Lüning U, Baumstark R, Wangnick C, Müller M, Schyja W, Gierst M, Gielbert M (1993) Pure Appl Chem 65: 527
9. Vinod TK, Hart H (1990) J Org Chem 55: 881
10. Grewal RS, Hart H, Vinod TK (1991) J Org Chem 57: 2721
11. Vinod TK, Hart H (unpublished results)
12. Hart H, Harada K (1985) Tetrahedron Lett 26: 46
13. Hart H, Harada K, Du C-JF (1985) J Org Chem 50: 3104
14. Vinod TK, Hart H (1988) Tetrahedron Lett 29: 885
15. Vinod TK, Rajakumar P, Hart H (unpublished results)
16. Grewal RS, Hart H (1990) Tetrahedron Lett 31: 4271
17. Vinod TK, Hart H (1991) J Org Chem 56: 5630
18. Lüning U, Müller M (1989) Chem Ber 123: 643; see also Lüning U, Baumstark R, Schyja W (1993) Tetrahedron Lett 34: 5063
19. Rajakumar P., Hart H (unpublished results; manuscript in preparation) see also Rajakumar P, Kannan A (1993) Tetrahedron Lett 34: 4407
20. Lüning U, Wangnick C, Peters K, von Schnering, HG (1991) Chem Ber 124: 397
21. Lüning U, Baumgartner H (1993) Syn Lett 8: 571

22. Ashok A, Hart H (unpublished results)
23. Rossa L. Vögtle F (1983) Top Curr Chem 113: 1
24. Anker W, Beveridge KA, Bushnell GW, Mitchell RH (1984) Can J Chem 62: 661
25. Mitchell RH (1983) In: Keehn PM, Rosenfeld SM (eds) Cyclophanes, vol 1. Academic Press, New York, p 240
26. Chiu J-J, Grewal RS, Hart H, Ward DL (1992) J Org Chem 58: 1553
27. Lüning U, Wangnick C (1992) Liebigs Ann Chem 481
28. Connolly routine is carried out with the QCPE program No 429 by ML Connolly, used with Chem-X, developed and distributed by Chemical Design Ltd. Oxford, England
29. Lüning U, Wangnick C (1992) Liebigs Ann Chem 481
30. Campbell M, Unrau C, Cox P, Snieckus V (Aug 26-31, 1990) Abstracts, 200th ACS meeting, Washington, DC, ORGN 187
31. Lüning U, Müller M (1992) Angew Chem Int Ed Engl 31: 80
32. Lüning U, Baumstark R, Müller M, Wangnick C, Schillinger F (1990) Chem Ber 221
33. Vinod TK, Hart H (1990) J Org Chem 55: 5461
34. Dijkstra G, Kruizinga WM, Kellog RM (1987) J Org Chem 52: 4230
35. Trost BM, Arndt HC, Stregl PE, Verhoeven TR (1976) Tetrahedron Lett 3477
36. Fujita T, Lehn JM (1988) Tetrahedron Lett 29: 1709
37. Dohm J, Vögtle F (1992) Top Curr Chem 161: 71
38. Umemoto T, Otsubo T, Misumi S (1974) Tetrahedron Lett 1573
39. Collman JP, Brauman JI, Fitzgerald JP, Hampton PD, Naruta, Y, Sparapany JW, Ibers JA (1988) J Am Chem Soc 110: 3477
40. a. Rubin Y, Dick K, Diederich F, Georgiadis TM (1988) J Org Chem 51: 3270; b. Loncharich RJ, Seward E, Ferguson SB, Brown FK, Diederich F, Houk KN (1988) J Org Chem 53: 3479; c. Miller SP, Whitlock HW, Jr (1984) J Am Chem Soc 106: 1492

Belt-, Ball-, and Tube-Shaped Molecules

Axel Schröder, Hans-Bernhard Mekelburger and Fritz Vögtle

Institut für Organische Chemie und Biochemie der Universität Bonn,
Gerhard-Domagk-Str. 1, 53121 Bonn, FRG

Table of Contents

This contribution gives a survey of recent developments in the field of belt-, ball-, and tube-shaped molecules. During the past few years there have been several break-throughs in this area. Therefore the development will progress much faster in the future than in the past. All new compounds discussed in this review contain cavities, holes, niches, or channels which are of topical interest in the fields of supramolecular chemistry and molecular recognition, as well as in molecular self-organization and sensor applications.

Topic in Current Chemistry, Vol. 172
© Springer-Verlag Berlin Heidelberg 1994

1 Introduction

In organic chemistry routes to "extraordinary" compounds have been looked for a long time. Molecules with familiar macroscopic structures (e.g. tetrahedron, cube, dodecahedron, football, crown, pagoda, cage, ...) belong to this group, too. Often molecular symmetry, steric strain and intra- or intermolecular interactions have been in the foreground.

Frequently, chemists chose trivial names for these fascinating molecules which on the one hand are pithier and shorter than the corresponding systematic terms and on the other hand also express the affinity of the molecules with distinct objects. The similarity of molecular and macroscopic structure can refer to the external shape, a special physical or chemical property or even both.

Molecular belts, molecular knots and trefoil knots, propellanes and pagodanes, even such a unique molecule like footballene or buckminsterfullerene belong to these compounds which remind the observer of a comparable macroscopic object [1].

In many cases very symmetrical molecules are concerned. Their structure is not sufficiently described any more by determination of the constitution (connectivity of atoms) and the Euclidean geometry (spatial arrangement of atoms because of molecular rigidity). For those structures an additional definition of isomerism is needed which exceeds constitutional and Euclidean stereoisomerism.

2 Topological Stereochemistry

The mathematical term *topology* involves geometrical properties which remain invariant given continuous deformation in space [2, 3]. Two structures are said to be *topological equivalent* or *isotopic* if they can be interconverted into one another by continuous deformation. If not they are called *topologically distinct* or *not isotopic*. One of these topological invariants is connectivity. Structures with identical connectivity are termed *homeomorphic*. Transferring these terms to molecular structure results in three types of isomers:

a) Constitutional isomers: differ in bond connectivity
   not homeomorphic, not isotopic
b) Euclidean stereoisomers: distinct by some kind of molecular rigidity
   homeomorphic, isotopic
c) Topological stereoisomers: distinct not because of molecular rigidity
   homeomorphic, not isotopic

The structures **XI** to **XIII** or **XIV** and **XV** are topological stereoisomers since they have identical connectivity (homeomorphic), but no continuous deformation will allow them to interconvert (not isotopic). Furthermore, structures **XII** and **XIII** are *topological enantiomers* and the knots **XII** or **XIII** and the unknotted ring **XI** are *topological diastereomers*.

The examples show that these novel molecular entities are not only fascinating, but that they should have interesting chemical and physical properties. For thirty

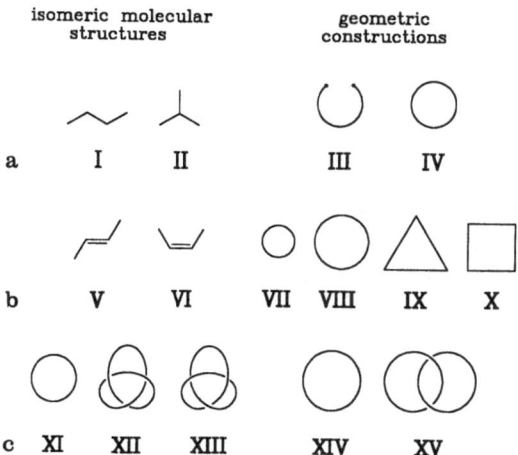

years chemists have tried to synthesize molecules which are similar to structures cited above and should show topological stereoisomerism. Until now the syntheses of catenanes, and even the first molecular Möbius strip have been successful [4].

In the following chapters a survey on belt-shaped molecules will point out that these molecules are fascinating compounds, however, their synthesis often requires a large amount of synthetic expenditure. The analysis of their features reveals that they deserve attention not only because of their unique structure and symmetry but also because of their properties which awaken interest in them, e.g. in supramolecular chemistry. The molecular structure gives rise to the presumption that they can act as novel hosts for ions and molecules or represent model substances for molecular pores, tubes or channels.

3 Cyclacenes, Beltenes and Collarenes

In 1983, Vögtle [5] proposed the synthesis of **1**, a molecular belt with a completely aromatic carbon skeleton. The benzenoid nuclei are fused laterally in a polyacene-like manner. The compound which could be called "superacene" [5] is a representative of the [n]cyclacenes **2** which gained more and more interest in the following years.

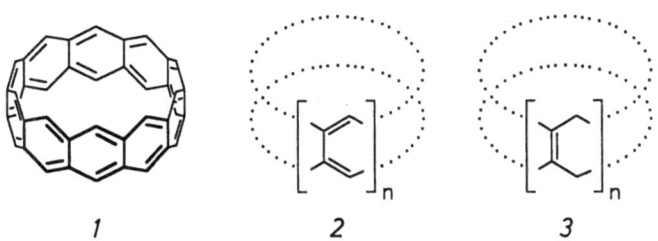

1   *2*   *3*

Thus in 1985, Alder and Sessions [6] carried out some force-field calculations on the [n]beltenes **3**. The authors realized, using models of the molecules, that [n]beltenes with n ≥ 6 span a cylindrical cavity which is suitable for including ions and small molecules. They also discussed the incorporation of quinone rings into the beltenes and postulated novel electron transfer reactions from the macrocyclic hosts to the complexed guests.

The calculations revealed that all [n]beltenes for n = 3 to 12 have an energy minimum in the configuration with D_{nh} symmetry, i.e. they have the form of a cylinder with a length of 700–800 nm. The strain energy per macro-ring unit decreases monotonically from [3]beltene to [12]beltene (about 25 kJ/mol). Host-guest complexes of the beltenes with acetylene should be favored energetically for n = 7 to 11 because of van der Waals interactions.

Angus und Johnson [7] published a theoretical paper in 1988. It deals with the π-π-interaction in the cavity of beltenes which they called "columnenes". The authors in accordance with Alder and Sessions predict a relatively low stability for [3]beltene. This is due to the extreme deformation of the π-orbitals arranged inwards. However, beltenes with n ≥ 4 there should be stable compounds which could be isolated.

Stoddart et al. [8] made a big step on the way to the first cyclacene. They succeeded in synthesizing the immediate precursor **6** by the Diels-Alder reaction of the two concave building blocks **4** and **5** under a pressure of 9–10 kbar.

Especially they took advantage of the endo-selectivity of the Diels-Alder reaction and the acceleration of rates by high pressures. The yield of the cyclic product **6**, a solid with a mp > 300 °C, increased to 20%. Good crystals for X-ray crystallography emerged from slow evaporation of a chloroform solution of **6**. The X-ray crystal structure discloses the elegance of the molecular structure and reveals the clathrate-like inclusion of chloroform molecules between layers of **6**.

In a further step, Stoddart et al. [9] tried to obtain [12]cyclacene and [12]beltene. By reaction of **6** with TiCl$_4$/LiAlH$_4$ in THF at ambient temperature two oxygen

atoms were removed. The macropolycycle **7**, mp > 250 °C, was isolated in 43%
yield. An X-ray crystal structure reveals that the rigid cavity has an approximately
square cross-section within which a water molecule is trapped.

The removal of the remaining four oxygen atoms was achieved by refluxing **7**
with a mixture of acetic anhydride and hydrochloric acid for 16 hours. But not
the "symmetrical" hydrocarbon **8** with two anthracene and two benzene units
was isolated. Dehydration was accompanied by some acid catalyzed isomerization
of the aromatic rings. The thermodynamically more stable but "unsymmetrical"
hydrocarbon **9** of low solubility and high thermal stability (mp > 600 °C) was
isolated in 56% yield.

The following Birch reduction, aimed at synthesizing [12]beltene, resulted in the
production of a symmetrical dodecahydro[12]cyclacene with the probable constitution **10** called [12]collarene by Stoddart. This structure with six isolated
benzene rings is believed to be the most stable compound along the route to
[12]beltene because of maximum resonance energy.

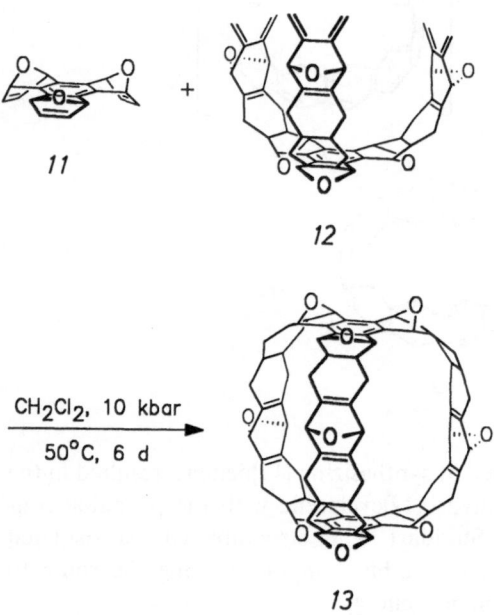

For **6** an interesting application has already been found [10]: If a molecule capable of complexing a definite guest is coated on the surface of a piezoelectric quartz crystal, the oscillation frequency of the crystal changes in the presence of the guest due to complexation and thus mass increase of the crystal. Measuring the change of frequency, the ratio between host and guest can be determined and possibly the concentration of guest in the gas phase could be measured which is important for trace analysis of organic molecules. In the case of the macropolycycle **6**, a high sensitivity for nitrobenzene was able to be demonstrated due to edge-to-face interactions.

In the following, Stoddart et al. [11] succeeded in synthesizing the even more sophisticated cage compound **13**, called "trinacrene". Starting with the trisdienophile **11** and the trisdiene **12** the product **13** was obtained by a threefold Diels-Alder reaction in dichloromethane under a pressure of 10 kbar. After chromatography, 2 mg equivalent to a yield of 4% of **13** was isolated (mp > 300 °C). The structure was determined by FAB mass spectrometry and ^{1}H-NMR spectroscopy.

The expenditure for the synthesis of **13** can be realized by the fact that the starting material **11** is obtainable in only 0.6% yield by reaction of hexabromobenzene with furan. Nevertheless the extension of the synthetic strategy of collarenes to macropolycyclic skeletons is a fascinating challenge.

Recently, Schlüter et al. [12] reported the synthesis of another beltene derivative. They tried to obtain belt-shaped polymers **16** by Diels-Alder polymerization of the monomer **14b** which contains a diene and a dienophile unit. As a "by-product" they isolated the [6]beltene derivative **15**. The yield was able to be increased to 70% in diluted solution. An X-ray crystal structure of the yellowish crystals, mp > 260 °C, indicated a somewhat flattened double-stranded cycle which spans only a small cavity. The intraanular distances between opposite six-membered rings are in the range of van der Waals distances.

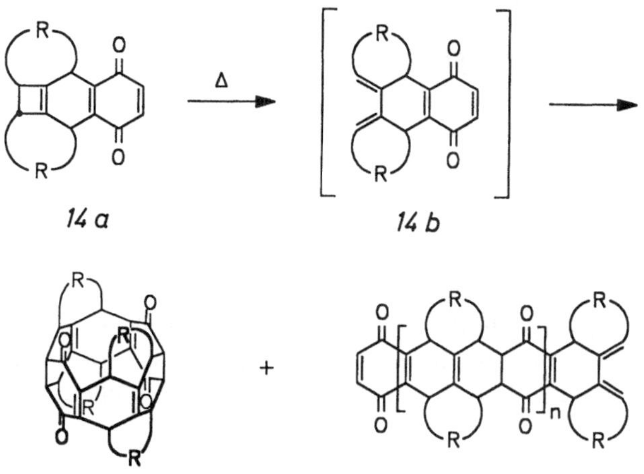

14 a   14 b

15   +   16

R = $-(CH_2)_6-$ n = 24–330

In this context it is worth mentioning the existance of hexahydro[3]beltene **17** which was synthesized as early as 1974 by Cupas et al. [13]. They named it "iceane" since its geometry is analogous to the crystal structure of water. **17** melts at 327–328.5 °C which is considerable for a saturated hydrocarbon with the empirical formula $C_{12}H_{18}$ (compare: adamantane melts at 269 °C).

17

4 π-Spherands

In 1977 and 1978, analyses of the complexing properties of crown derivatives with p-phenylene or 2,3-naphthyl units were carried out by several groups. They found out that the aromatic ring acts as a π-donor and thus contributes to the binding of the guest (usually an alkali metal cation) [14, 15, 16, 17, 18].

The first complex where the central atom is bound exclusively by π-orbital interactions was produced by Pierre et al. [19]. They treated [2.2.2]paracyclophane (18) with silver triflate and observed an interaction of the silver cation with the "π-prismand" which was one hundred times stronger than with open-chained aromatic compounds.

An X-ray crystal structure of the analogous silver perchlorate complex shows that the silver ion is not situated exactly in the center of the cavity but is displaced outwards for about 240 pm [20]. Boekelheide et al. [21] confirmed this phenomenon by carrying out an X-ray crystal structure of the silver triflate complex 19 where the silver ion is likewise not situated in the center between the three p-phenylene rings.

In 1985, Boekelheide et al. [21] succeeded in synthesizing the first belt-shaped molecule on the basis of orthocyclophanes. Benzo[1,2; 4,5]dicyclobutene (20) forms 21 as an intermediate during gas-phase pyrolysis at 425 °C in a stream of nitrogen. This intermediate 21 reacts with 20 to form the desired product 22 in 2.3% yield.

The compound 22 was named "deltaphane" because of its symmetric triangular structure or $[2_6](1,2,4,5)$cyclophane in cyclophane nomenclature [22]. An X-ray crystal structure of the pale yellow crystals which decompose above 345 °C elucidates the face-to-face arrangement of the three benzene rings in the symmetrical structure of the molecule. Good complexing properties are expected due to this rigid skeleton. Indeed the silver triflate complex 23, a pale yellow solid which decomposes above 330 °C, was prepared in 85% yield. The X-ray crystal structure reveals that analogous to the [2.2.2]paracyclophane, the silver ion is not situated in the center of the cavity. A comparison of the thermodynamic stability of the two complexes 19 and 23 by ^{1}H-NMR spectroscopy indicates that the silver complex of [2.2.2]paracyclophane is about 2.4 times more stable than the one of the "deltaphane". An explanation for this could be the larger flexibility of 18 which makes possible a better orbital interaction between the π-spherand and the included ion.

Another class of π-spherands with a belt-shaped structure originates from the connection of 1,4-cyclohexanediylidene units to macrocycles of type 24. However, in contrast to the beltenes the π-bonds are in the plane of the ring and not perpendicular to it. The most simple representative of this class (n = 2) is tricyclo[4.2.2.2^{2,5}]dodeca-1,5-diene (25) synthesized in 1981 by Wiberg et al. [23], but due to its small size it cannot be called a molecular belt.

24   25

Higher homologues with n = 3, 4, and 6 were prepared in the following by McMurry et al. with the aid of the synthetic method named after him. Slow addition of the open-chained diketone 26 to a boiling suspension of TiCl$_3$, zinc, and copper in THF gave the trimer 27, mp 259–259.5 °C, in 24% yield and only traces of the hexamer 28 [24].

26

27   28

An X-ray crystal structure proved the D_{3h} symmetry of **27**. The three six-membered rings exist in a boat conformation. **27** was expected to be homoaromatic but with the aid of photoelectron spectroscopy only a relatively small interaction between the π-fragments could be proved. The term homoaromaticity describes systems in which a stabilized cyclic conjugated system containing $(4n + 2)$ π-electrons is formed by bypassing one or more saturated atoms on the condition that a considerable interaction between the π-fragments exists due to the proximity of p-orbitals [25].

The synthesis of the macrocycle **30** with n = 4, pentacyclo[12.2.2.2^{2,5}.2^{6,9}.2^{10,13}]-tetracosa-1,5,9,13-tetraene, was carried out likewise by McMurry et al. [26]. They used the method cited above to cyclize the diketone **29** in an intramolecular reaction. The macrocycle **30**, mp > 300 °C, was isolated in 90% yield which is extremely high. An X-ray crystal structure showed that the distance between two opposite double bonds is 511 pm, i.e. the metal-carbon distance should be about 250 pm for optimal complexation of a metal cation. For this reason the silver ion was chosen. If the tetraalkene **30** is stirred with silver triflate in THF the expected complex **31**, stable against light, heat, and air, precipitates as a colorless solid in 66% yield (mp 145 °C).

The silver ion is situated exactly in the center of the cavity as an X-ray crystal structure demonstrates. Therefore **31** is the first square-planar d^{10}-organometallic complex. Besides, carbon-silver couplings in the ^{13}C-NMR spectrum prove that the nature of the complex is static, i.e. the central ion is bound tightly to the ligand **30**.

5 Crowns

In 1981 Walba et al. [27] developed a conception for the synthesis of a novel class of molecules composed of crown ether rings fused by the *tetrakis(hydroxymethyl)-ethylene* (THYME) unit. The target molecule **33** was designed as a cylindrical host with a hydrophilic interior surface − a three-dimensional analogue of the "flat" 18-crown-6. It was expected to have interesting properties and perhaps some catalytic activity, too. The key step on the route to **33** has been the preparation of the ladder-shaped diol-ditosylate **32** which was produced in several steps starting with a furan derivative. Cyclization of **32** to **33** was carried out under high dilution conditions in DMF using sodium hydride as an auxiliary base. The macrotricyclic polyether **33**, mp 118–123 °C, was obtained in 30–35% yield.

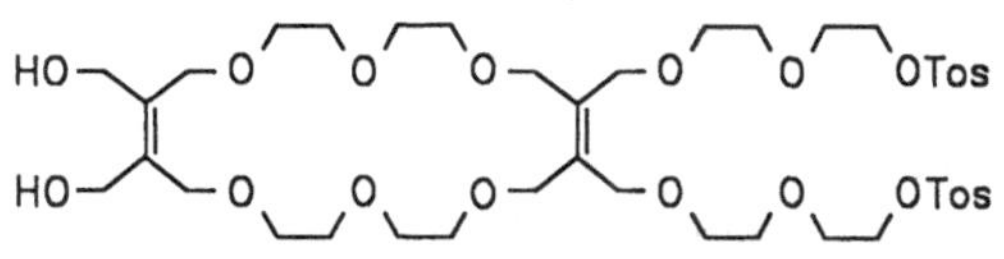

32

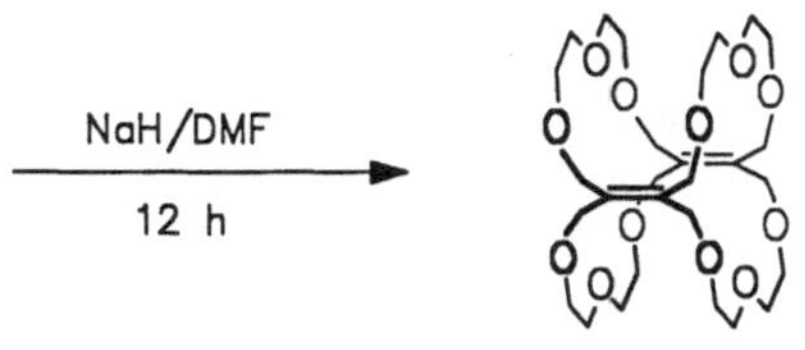

33

Recrystallization from toluene afforded crystals suitable for X-ray analysis which contained one molecule of the solvent per crown ether molecule. The free ligand adopts a conformation in the crystal which fills the cavity space almost completely. The toluene molecules are not strongly associated with those of the tricycle. This can be perceived by the easy loss of toluene from the solvate.

Crystals of the unsolvated macrocycle could be obtained by diffusion of hexane into a chloroform solution of **33** [28]. Even in this case the molecule assumes a "closed" conformation, i.e. the polyether chains approach each other very much. This observation corresponds with X-ray crystal structures of other crown ethers.

Treatment of ligand **33** with aqueous K$_2$PtCl$_4$ gave the tetrachloroplatinate complex **34**. Dissolution of complex **34** in DMSO resulted in the trinuclear cascade complex **35**. In this complex each of the two potassium ions is encapsulated by

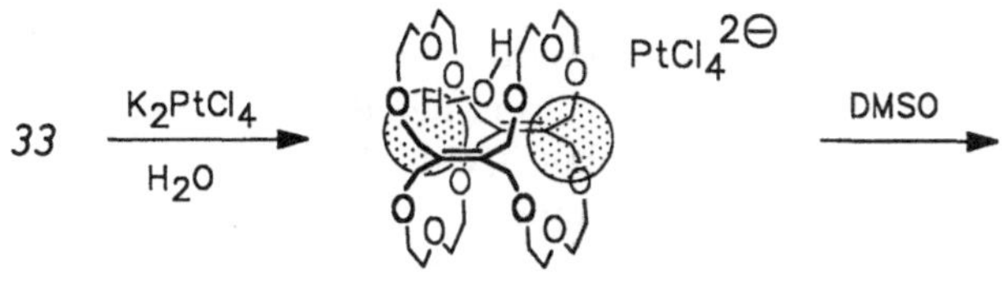

34

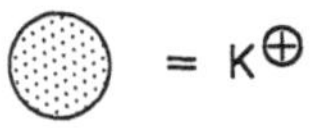

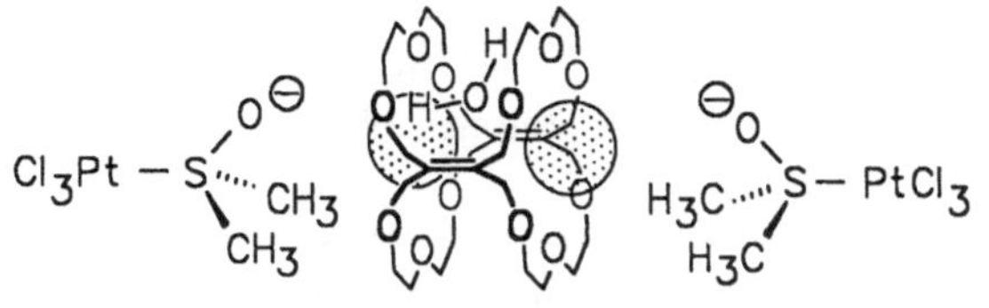

⬤ = K$^\oplus$

35

a 20-crown-6 moiety as an X-ray structure analysis shows. A $PtCl_3[DMSO]^-$ anion is associated with each potassium ion, bonded via the DMSO oxygen. The two potassium nuclei are separated by 491 pm which is an extraordinary close distance for two alkali-metal cations. In the absence of the ligand 33 the Coulombic repulsion between these two cations would be 280 kJ mol^{-1}.

Walba et al. extended their conception to the synthesis of belt-shaped crown ethers by lengthening the open-chained precursor 32 by one unit [29]. The cyclization of the diol-ditosylate 36 was carried out analogously to the method cited above and gave two products with the same molecular weight as determined by CI mass spectrometry in 50% overall yield. As CPK models suggest the four functional groups of 36 can react with each other not only "parallel" but also "crossed". This leads to the cylinder 37 and the first molecular Möbius strip 38. A Möbius strip as an isomer of a cylinder has only **one** surface and **one** edge.

36

NaH/DMF
rt, 48 h

37   38

Separation of both isomers succeeded by flash chromatography on alumina. Cylinder 37, mp 107–108 °C, was obtained in 24% and Möbius strip 38 (oil) in 22%. Whereas the exact constitution of 37 was proved by X-ray crystallography, the Möbius strip 38 could only be proved indirectly by its chirality. The compound could not be separated into its enantiomers, but ^{13}C-NMR spectroscopy in the presence of optically active Pirkle's reagent showed a splitting of the signal of the olefinic carbon atoms. The chirality of 38 represents a novel kind of optical isomerism called *topological enantiomerism* since neither an asymmetric center nor a rigid skeleton are necessary to distinguish between both enantiomers.

A second possibility of proving the structure of 38 consisted in the cleavage of the double bonds, the "rungs" of the Möbius-"ladder", by ozonolysis [30]. While the cylinder 37 led to two 30-membered cyclic triketones 39, the Möbius strip 38 gave a 60-membered cyclic hexaketone 40.

In a subsequent step, Walba et al. [30] attempted to synthesize the more highly twisted products 42 and 43 by using the once more lengthened, now 4-rung, ladder-shaped diol-ditosylate 41 as precursor. But again the cylinder and the simple

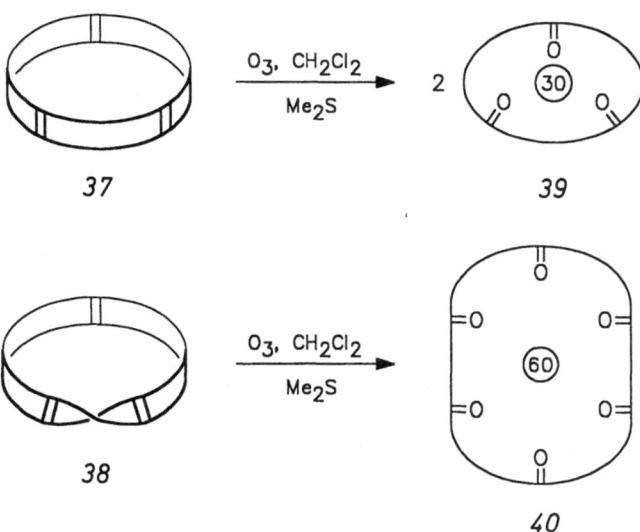

Möbius strip were formed exclusively. The synthesis of the more highly twisted belts **42** and **43** would be very interesting because subsequent cleavage of the double bonds by ozonolysis would lead to a catenane **44** and a molecular trefoil knot **45**, respectively.

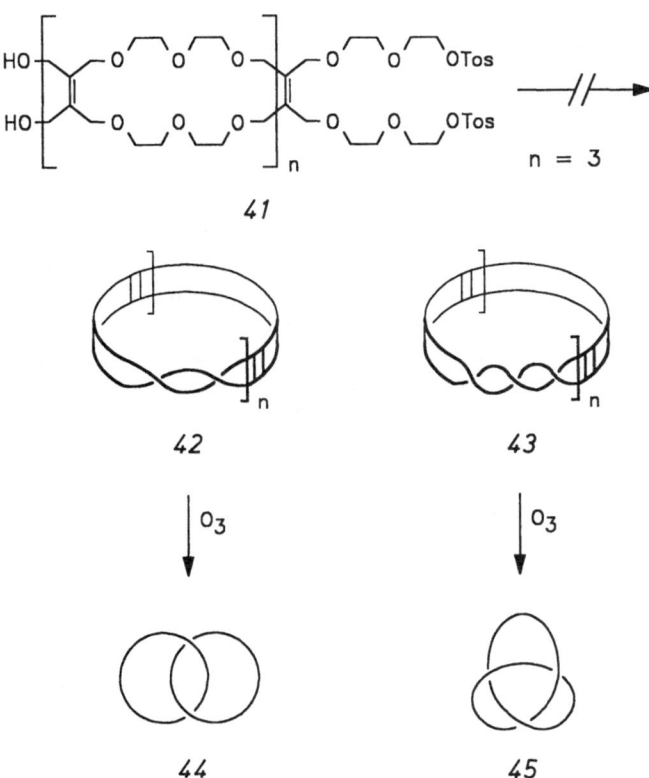

6 Cavitands and Carcerands

Molecules which do not have a cylindrical but a conical cavity belong to the group of belt-shaped molecules, too. Among them are the "cavitands" **46** described by Cram et al. [31]. Usually they contain four aromatic rings which are combined to a macrocycle by *ortho*-connection. However, the two bridges which are in *ortho*-position have different lengths, so that the aromatic rings do not lie on the surface of a cylinder but on the surface of a cone. Synthetic precursors of the cavitands are calixarenes substituted by phenolic OH-groups with the general structure **47** [32, 33, 34, 35]. Discovered as early as 1872 by A. v. Baeyer [36], they are condensation products between phenols and formaldehyde. They are easily accessible because of their simple syntheses. Up to now a large number of stable and crystalline complexes of different calixarenes with small molecules (mostly solvent molecules) have been isolated.

$$X = SiR_2, (CR_2)_n$$

The cavitands **46** in comparison with the calixarenes **47** have a more rigid skeleton and therefore they are conformationally less flexible due to intensified bridging of their aromatic moieties. Molecules with a relatively rigid structure are, because of their superior preorganization (i.e. sterical fixation of possible binding sites), usually more capable of complexation compared to molecules which can exist in many different stable conformations [37]. Varying the substituents on the skeleton of the cavitands and varying the length of the bridges between the benzene units, Cram et al. [38] were able to change the shape and depth of the cavity in such a way that molecules of different sizes like dichloromethane, trichloromethane, cyclohexane, benzene, or toluene were able to be complexed because of the complementarity of host and guest. The structure of all complexes is demonstrated by X-ray crystallography. In almost all cases the guest molecule is situated in the cavity of the host.

Cavitands can be functionalized on four positions very easily. Cram et al. [39, 40] took advantage of this fact and coupled two of these functionalized bowl-shaped building blocks, **48** and **49**, under high dilution conditions to the cage-like compound **50** called "carcerand". This name was chosen due to the compact structure of these macropolycycles which are capable of imprisoning included guest molecules and even atoms as if in a jail ("carcer") since the size of the holes is not big enough for decomplexation.

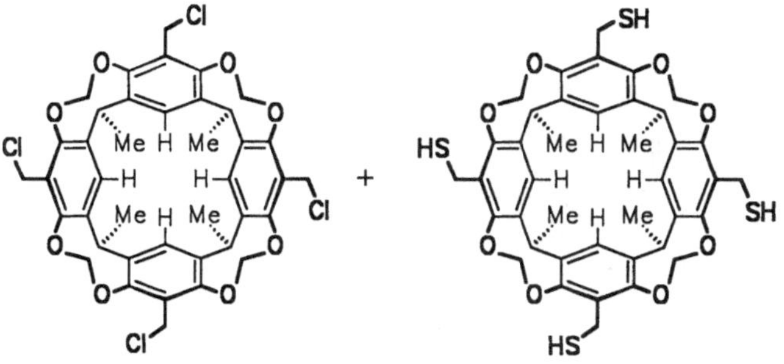

48   +   *49*

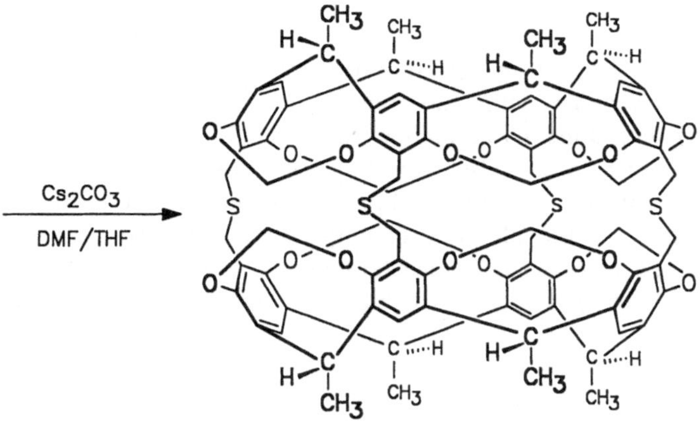

50

Indeed, after analyzing the product of the cyclization it was found that, e.g. solvent molecules like DMF but even cesium ions or argon atoms were present as permanent guests in the cavity of the carcerand. This fact was proved by FAB mass spectrometry and elemental analyses. The crude mixture of carcerand **50** and carceplexes **50 · G** (G stands for incarcerated guest entities) obtained in a yield of 29% proved to be essentially insoluble. This prohibited further purification and prevented the determination of its ^{1}H-NMR spectrum. A solid state ^{13}C-NMR spectrum was determined instead which showed the expected eight different signals.

Carcerand **51** is much more soluble than **50** due to eight 2-phenylethyl substituents [41]. During its synthesis a template effect by various solvent molecules was observed. Thus cyclizing the precursor in dimethyl sulfoxide, dimethyl acetamide, and dimethyl formamide gave the corresponding carceplexes **51 · G** in 61%, 54%, and 49% yields, respectively (i.e. the interior of the carcerands was occupied by one solvent molecule).

51

R = CH$_2$CH$_2$Ph

If the cyclization was carried out in a mixture of *N*-formylpiperidine (which is too large for incarceration) and dimethyl acetamide in a ratio of 99.5 to 0.5 only the 1 : 1 complex of **51** with dimethyl acetamide was isolated in 10% yield. The encapsulation of guest molecules was also proved by ^{1}H-NMR spectroscopy. All protons of the guests are shifted upfield by 1 to 4 ppm from their normal positions. These large shifts are realized by the enforced proximity to the arene shielding zone.

The authors designate carcerands as "molecular container compounds" due to their property of not releasing included particles any more. They even use the term "molecular cell" [40].

7 Cucurbituril

Another belt-shaped molecule is cucurbituril (**53**) which was presumably synthesized as early as 1905 by Behrend et al. [42]. The structural formula of **53** was resolved in 1981 by Freeman, Mock, and Shih [43] only when modern spectroscopic methods were available. Cucurbituril (**53**) is easily obtained from urea, glyoxal, and formaldehyde in an acid-catalyzed condensation reaction with glycoluril (**52**) as an intermediate. The initially formed polycondensation product, insoluble in all common solvents, is treated with hot concentrated sulfuric acid. After dilution with water and subsequent boiling a solid is obtained in a yield of 40–70%. The ^{1}H-NMR spectrum shows only three signals of equal intensity and the infrared spectrum suggests retention of the glycoluril moieties. The compound is not suitable for usual mass spectrometry due to its low volatility. However, an X-ray crystal structure of the calcium complex obtained from sulfuric acid solution was undertaken.

It revealed that **53** is a cyclic hexamer of dimethanoglycoluril with a relatively rigid skeleton and an internal cavity of approximately 550 pm diameter. Both

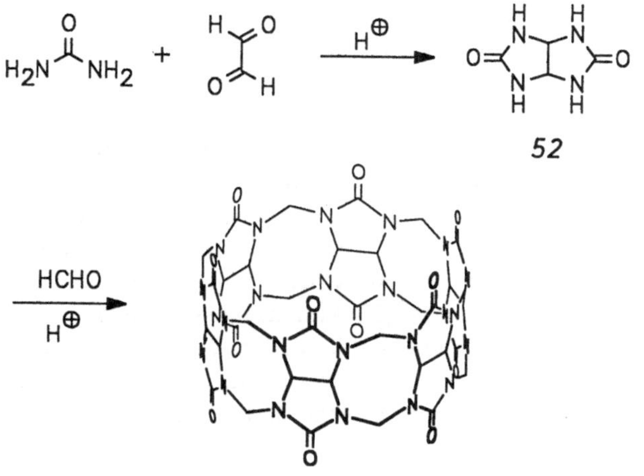

52

53

openings of the macrocycle have a diameter of 400 pm. The center of each hexamer is occupied by a free water molecule. The calcium ions are coordinated to the oxygen atoms of the urea carbonyl groups and in addition to water and sulfate ligands.

Complexation experiments by Mock et al. [44, 45, 46] proved that Cucurbituril (53) forms 1 : 1 complexes with a number of alkylammonium salts where the guest is situated in the cavity of the host. The most stable host-guest complexes are observed with alkyldiammonium salts; the dissociation constants are remarkably high: 10^{-6} to 10^{-7} mol l^{-1}. Binding results mainly from ion-dipole interactions and hydrogen bonding between the ammonium groups and the carbonyl oxygen atoms of the host. The hydrocarbon chain of the guest is situated in the cavity favoured by the hydrophobic effect. This is proved by ^{1}H-NMR spectroscopy. All alkyl protons show upfield shifts of up to 1 ppm. Ammonium salts with bulkier hydrocarbon moieties (i.e. tolyl or cyclohexyl groups) are discriminated since they do not fit into the cavity of **53** due to their size.

Mock et al. [47] used the complexation of alkylammonium ions by Cucurbituril (53) to catalyze 1,3-dipolar cycloadditions of ammonium substituted alkynes **54** and alkyl azides **55**. This pericyclic reaction is accelerated by a factor of 5.5×10^4 under the catalytic influence of **53**.

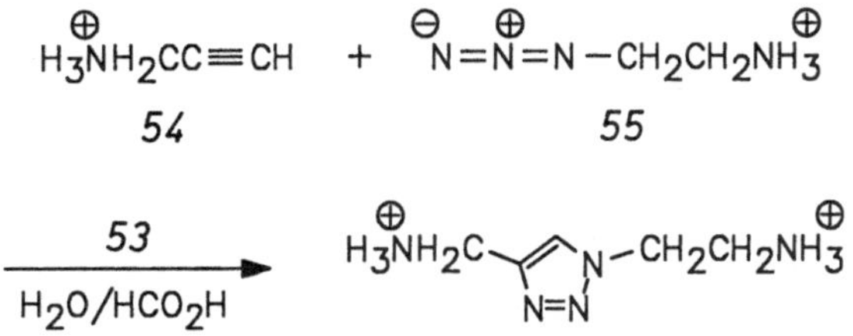

54   *55*

56

In this case, Cucurbituril (**53**) reveals a number of enzymelike features: The reaction exhibits saturation behaviour, it becomes independent of substrate concentration with sufficient amounts of **54** and **55**, high concentrations of **54** retard the cycloaddition (substrate inhibition), and release of product **56** from its complex with Cucurbituril (**53**) is the rate determining step. NMR spectroscopic data suggest that both starting materials of the cycloaddition are hydrogen bonded to the carbonyl groups of **53** with their ammonium moiety and that the reactive substituents extend into the interior of Cucurbituril (**53**). In this cavity the pericyclic reaction takes place to form the 1,2,3-triazole **56**. Kinetic data indicate that the formation of the ternary complex of Cucurbituril (**53**) with the two starting materials **54** and **55** is not strainless. Since the reaction is still accelerated very much it is assumed that the transition state of the reaction corresponds to the size of the cavity more closely than the substrates. This is a further indication that this case is a useful enzyme model.

8 Ball-shaped Molecules: C_{60} and C_{70}, two new Allotropic Forms of Carbon

At this point two molecules will be discussed which strictly speaking do not have a belt- or tube-shaped structure but do have subunits with these structures. Doubtlessly C_{60} (**57**) and C_{70} (**58**) are highlights in recent organic chemistry. Their existence was proved in 1990 after a long period of speculation [48].

As early as 1985, Kroto, Smalley et al. [49] reported on the observation of a stable C_{60} particle by time-of-flight mass spectrometry following vaporization of graphite by irradiation with a laser beam. They proposed structure **57** for this carbon cluster which resembles a soccer ball. Thus the trivial name footballene or buckminsterfullerene was given to **57**. It consists of five- and six-membered rings only which are fused to a ball-shaped skeleton with an internal cavity. After impregnation of the graphite rod with lanthanum C_nLa-complexes could be observed mass spectrometrically with the $C_{60}La$-complex as the main product [50]. In these complexes one single lanthanum atom is presumably sheathed by an aromatic carbon shell. The same authors postulated the oval C_{70} (**58**) as a further stable cluster with a spherical carbon shell, consisting solely of six-membered rings.

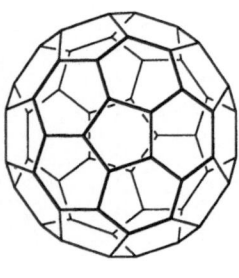

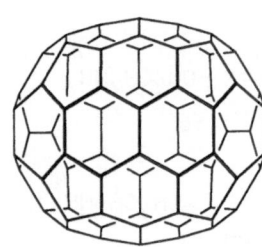

57   *58*

A confirmation of the prediction referring to the molecular structure of C_{60} and C_{70} was given in 1990 by Krätschmer et al. [51]. They synthesized and purified 100 mg amounts of the compounds concerned by resistively heating graphite in a helium atmosphere at about 100 Torr and extracting the raw carbon dust with hot benzene. Separation of C_{60} from C_{70} was achieved by chromatography or by fractional sublimation at very high temperatures by various groups [48]. It was shown that the ratio between C_{60} and C_{70} is about 85:15. The amounts obtained were sufficient to get ^{13}C-NMR spectra (in benzene) of both molecules. For C_{60} only one signal is expected because of its high symmetry. This is confirmed experimentally [52, 53, 54, 55]. This single signal is observed at $\delta = 143.2$ which is in the range of aromatic carbon atoms. Thus this compound is the third "member" of the point group with icosahedral symmetry (I_h) until now after the borohydride anion ($B_{12}H_{12}^{2-}$) and dodecahedrane ($C_{20}H_{20}$). The molecular structure of C_{70} was also proved by its ^{13}C-NMR spectrum [56, 57]. Five signals between 130 and 150 ppm were observed which are consistent with a D_{5h} symmetry.

The fact that these two new carbon modifications are stable compounds with which any chemistry can be done was shown by Smalley et al. [58]. They found out that C_{60} undergoes a Birch reduction to a mixture of isomers with the empirical formula $C_{60}H_{36}$. After subsequent dehydrogenation with 2,3-dichloro-5,6-dicyano-1,4-benzoquinone (DDQ) C_{60} is recovered again. This indicates that the reduction is fully reversible and that the skeleton of the molecule is not altered during the reaction.

The relation between these two very interesting structures and belt- or tube-shaped molecules becomes clear after careful contemplation of C_{70} (**58**). One recognizes an aromatic belt consisting of benzene and naphthalene rings as a subunit which very much resembles molecule **59** proposed by Vögtle [5].

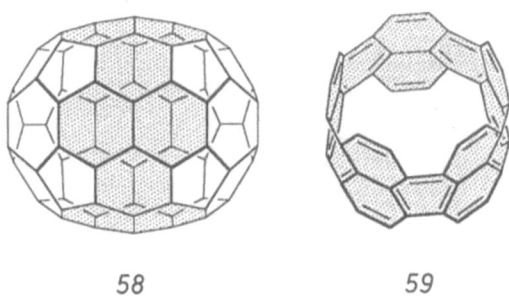

58   59

9 Tube-Shaped Molecules

Recently, Vögtle et al. [59] synthesized the first tube-shaped molecule based on metacyclophanes. In this case the key building block **61**, a four-fold functionalized diaza[3.3]metacyclophane, was accessible by reaction of **60** with the mono sodium salt of 4-toluenesulfonamide. The conversion of the peripheral functional groups was achieved easily due to the stable N-Tos bridges.

In this way compounds **62** and **63** required for the final cyclization were synthesized. This cyclization to the tube-shaped macropolycycle with four aromatic subunits was carried out using the cesium effect [60] and the dilution principle [61]. Even upon careful work-up by means of HPLC only one of the two conceivable isomers **64** and **65** was isolated.

On the basis of ^{1}H-NMR spectroscopical data, especially by taking into consideration the influence of the magnetic anisotropism of the neighboring benzene ring on the intraanular hydrogen atoms of the cyclophane units, structure **64** was assigned to the isolated compound. Comparisons with known cyclophanes were also carried out. In this context the temperature dependence of the ^{1}H-NMR spectrum was decisive. The reason for this phenomenon is a relative slow transformation of conformation **64a** to conformation **64b** and vice versa at room temperature. Therefore all signals are doubled which results in more information with regard to the structure.

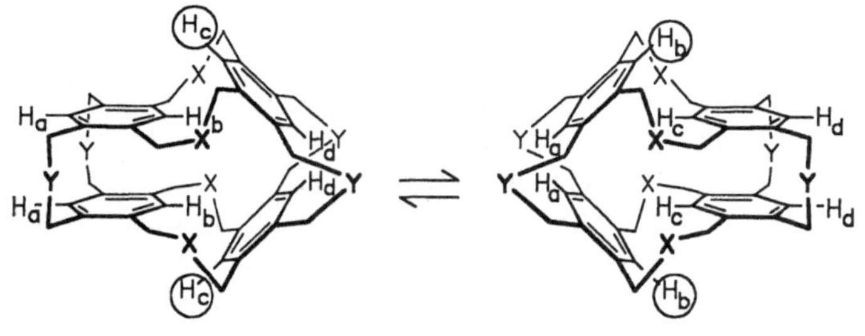

X,Y = S, N—Tos, resp.

64a   *64b*

Noteworthy in the ^{1}H-NMR spectrum of **64** at ambient temperature is the signal at $\delta = 8.75$ which has an extreme downfield position. This signal can only be attributed to the hydrogen atoms H_c and H_b, respectively. All other hydrogen atoms should absorb at higher field. At 160 °C a ^{1}H-NMR spectrum is obtained where the signals are not doubled any more due to faster transformation of the

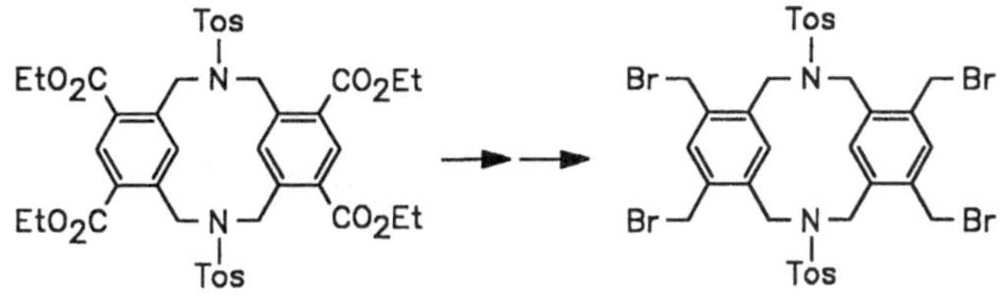

61   *62*

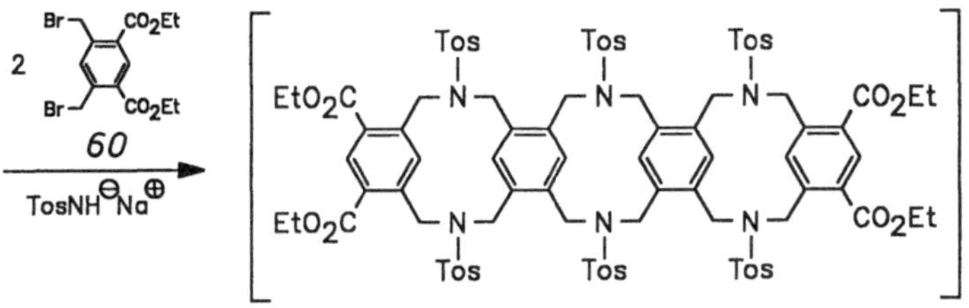

66

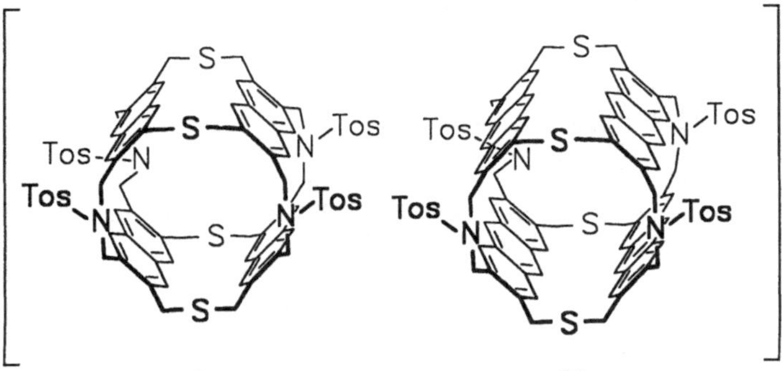

67   *68*

conformations. The energy barrier for this process is approx 65–70 kJ mol^{-1}, the coalescence temperature is approx 70 °C (calculated for H_b and H_c with $\Delta v \approx 215$ Hz).

The synthetic strategy applied here is generally applicable and gives access to a number of similar molecular structures. This was shown recently by the cyclization of **62** with 3,3',5,5'-tetrakis(mercaptomethyl)biphenyl which was proved mass spectrometrically [62]. On the one hand the repetitive prolongation of precursors (e.g. **61** to **66**) leads to tube-shaped molecules with a larger diameter, on the other hand the tube-length can be varied, e.g. by use of naphthalenophanes or anthracenophanes (**67** and **68**). Thus in the future it should be possible to prepare tube-shaped "endoreceptors" (longer, expanded ring hydrocarbons, Möbius strips). Because of their channel-type cavities they should also be of interest in supramolecular chemistry.

As one can see, molecules which at first sight seem to be utopian can be synthesized although isolation very often needs a lot of effort. Therefore they are still rewarding goals and interesting subjects for chemists. In particular for tube-shaped molecules the future potentialities are of great variety.

10 References

1. Vögtle F (1989) Reizvolle Moleküle der Organischen Chemie, Teubner, Stuttgart; (1992) Fascinating molecules in organic Chemistry, Wiley, Chichester
2. Frisch HL, Wassermann E (1961) J Am Chem Soc 83: 3789
3. Walba DM (1985) Tetrahedron 41: 3161
4. Vögtle F (1989) Supramolekulare Chemie, Teubner, Stuttgart; (1991) Supramolecular chemistry, Wiley, Chichester
5. Vögtle F (1983) Top Curr Chem 115: 157
6. Alder RW, Session RB (1985) J Chem Soc, Perkin Trans II 1849
7. Angus Jr RO, Johnson RP (1988) J Org Chem 53: 314
8. Kohnke FH, Slawin AMZ, Stoddart JF, Williams DJ (1987) Angew Chem 99: 941; (1987) Angew Chem, Int Ed Engl 26: 892
9. Ashton PR, Isaacs NS, Kohnke FH, Slawin AMZ, Spencer CM, Stoddart JF, Williams DJ (1988) Angew Chem 100: 981; (1988) Angew Chem, Int Ed Engl 27: 966
10. Elmosalamy MAF, Moody GJ, Thomas JDR, Kohnke FH, Stoddart JF (1989) Anal Proc London 26: 12
11. Ashton PR, Isaacs NS, Kohnke FH, d'Alcontres GS, Stoddart JF (1989) Angew Chem 101: 1269; (1989) Angew Chem, Int Ed Engl 28: 1261
12. Godt A, Enkelmann V, Schlüter AD (1989) Angew Chem 101: 1704, (1989) Angew Chem, Int Ed Engl 28: 1680
13. Cupas CA, Hodakowski L (1974) J Am Chem Soc 96: 4668
14. Wester N, Vögtle F (1978) J Chem Res (S) 400
15. Frensch K, Vögtle F (1977) Tetrahedron Lett 30: 2573
16. Beckford HF, King RM, Stoddart JF (1978) Tetrahedron Lett 31: 171
17. Kawashima N, Kawashima T, Otsubo T, Misumi S (1978) Tetrahedron Lett 31: 5025
18. Sousa LR, Larson JM (1977) J Am Chem Soc 99: 307
19. Pierre JL, Baret P, Chautemps P, Armand M (1981) J Am Chem Soc 103: 2986
20. Cohen-Addad C, Baret P, Chautemps P, Pierre JL (1983) Acta Crystallogr, Sect C 39: 1346
21. Kang HC, Hanson AW, Eaton B, Boekelheide V (1985) J Am Chem Soc 107: 1979
22. Boekelheide V (1983) Top Curr Chem 113: 89
23. Wiberg KB, Matturro M, Adams R (1981) J Am Chem Soc 103: 1600

24. McMurry JE, Haley GJ, Matz JR, Clardy JC, Van Duyne G (1984) J Am Chem Soc 106: 5018
25. Martin HD, Mayer B (1983) Angew Chem 95: 281; (1983) Angew Chem, Int Ed Engl 22: 283
26. McMurry JE, Haley GJ, Matz JR, Clardy JC, Mitchell J (1986) J Am Chem Soc 108: 515
27. Walba DM, Richards RM, Sherwood SP, Haltiwanger RC (1981) J Am Chem Soc 103: 6213
28. Walba DM, Richards RM, Hermsmeier M, Haltiwanger RC (1987) J Am Chem Soc 109: 7081
29. Walba DM, Richards RM, Haltiwanger RC (1982) J Am Chem Soc 104: 3219
30. Walba DM, Armstrong J, Perry AE, Richards RM, Homan TC, Haltiwanger RC (1986) Tetrahedron 42: 1883
31. Cram DJ (1983) Science 219: 1177
32. Zinke A, Zigeuner G (1944) Ber Dtsch Chem Ges 77B: 264
33. Gutsche CD (1983) Acc Chem Res 16: 161
34. Gutsche CD (1990) In: Stoddart JF (ed) Calixarenes, Royal Society, London
35. Vögtle F (1990) Cyclophan-Chemie, Teubner, Stuttgart
36. von Baeyer A (1872) Ber Dtsch Chem Ges 1094
37. Cram DJ (1986) Angew Chem 98: 1041; (1986) Angew Chem, Int Ed Engl 25: 1039
38. Cram DJ, Karbach S, Kim HE, Knobler CB, Maverick EF, Ericson JL, Helgeson RC (1988) J Am Chem Soc 110: 2229
39. Cram DJ, Karbach S, Kim YH, Baczynskyj L, Kalleymeyn GW (1985) J Am Chem Soc 107: 2575
40. Cram DJ, Karbach S, Kim YH, Baczynskyj L, Marti K, Sampson RM, Kalleymeyn GW (1988) J Am Chem Soc 110: 2554
41. Sherman JC, Cram DJ (1989) J Am Chem Soc 111: 4527
42. Behrend R, Meyer E, Rusche F (1905) Liebigs Ann Chem 339: 1
43. Freeman WA, Mock WL, Shih NY (1981) J Am Chem Soc 103: 7367
44. Mock WL, Shih NY (1983) J Org Chem 48: 3618
45. Mock WL, Shih NY (1986) J Org Chem 51: 4440
46. Mock WL, Shih NY (1988) J Am Chem Soc 110: 4706
47. Mock WL, Irra TA, Wepsiec JP, Mavimaran TL (1983) J Org Chem 48: 3619
48. Stoddart JF (1991) Angew Chem 103: 71; (1991) Angew Chem, Int Ed Engl 30: 70
49. Kroto HW, Heath JR, O'Brien SCO, Curl RF, Smalley RE (1985) Nature 318: 162
50. Heath JR, O'Brien SCO, Zhang Q, Liu Y, Curl RF, Kroto HW, Tittel FK, Smalley RE (1985) J Am Chem Soc 107: 7779
51. Krätschmer W, Lamb LD, Fostiropoulos K, Huffman DR (1990) Nature 347: 354
52. Taylor R, Hare JP, Abdul-Sada AK, Kroto HW (1990) J Chem Soc, Chem Commun 1423
53. Johnson RD, Meijer G, Bethune DS (1990) J Am Chem Soc 112: 8983
54. Yannoni CS, Bernier PP, Bethune DS, Meijer G, Salem JR (1991) J Am Chem Soc 113: 3190
55. Allemand PM, Srdanov G, Koch A, Khemani K, Wudl F, Rubin Y, Diederich F, Alvarez MM, Anz SJ, Whetten RL (1991) J Am Chem Soc 113: 2780
56. Ajie H, Alvarez MM, Anz SJ, Beck RD, Diederich F, Fostiropoulos K, Huffman DR, Krätschmer W, Rubin Y, Schriver KE, Sensharma D, Whetten RL (1990) J Phys Chem 94: 8630
57. Johnson RD, Meijer G, Salem JR, Bethune DS (1991) J Am Chem Soc 113: 3619
58. Haufler RE, Conceicao J. Chibante LPF, Chai Y, Byrne NE, Flanagan S, Haley MM, O'Brien SCO, Pan C, Xiao Z, Billups WE, Ciufolini MA, Hauge RH, Margrave JL, Wilson LJ, Curl RF, Smalley RE (1990) J Phys Chem 94: 8634
59. Vögtle F, Schröder A, Karbach D (1991) Angew Chem 103: 582; (1991) Angew Chem, Int Ed Engl 30: 575
60. Ostrowicki, A, Koepp E, Vögtle F (1991) Top Curr Chem 161: 37
61. Knops P, Sendhoff N, Mekelburger HB, Vögtle F (1991) Top Curr Chem 161: 1
62. We kindly thank Dipl.-Chem. R. Güther for the preparation of 3,3',5,5'-tetrakis(mercaptomethyl)biphenyl: Güther R, (1990) Diplomarbeit, Univ Bonn

201

Author Index Volumes 151-172

The volume numbers are printed in italics

Stein, N., see Bley, K.: *166*, 199-233 (1993).
Stoddart, J.F., see Kohnke, F.H.: *165*, 1-69 (1993).
Soumillion, J.-P.: Photoinduced Electron Transfer Employing Organic Anions. *168*, 93-141 (1993).
Stumpe, R., see Kim, J.I.: *157*, 129-180 (1990).
Suami, T.: Chemistry of Pseudo-sugars. *154*, 257-283 (1990).
Suppan, P.: The Marcus Inverted Region. *163*, 95-130 (1992).
Suzuki, N.: Radiometric Determination of Trace Elements. *157*, 35-56 (1990).

Thiem, J., and Klaffke, W.: Synthesis of Deoxy Oligosaccharides. *154*, 285-332 (1990).
Timpe, H.-J.: Photoinduced Electron Transfer Polymerization. *156*, 167-198 (1990).
Tobe, Y.: Strained [n]Cyclophanes. *172*, 1-40 (1994.
Tolentino, H., see Fontaine, A.: *151*, 179-203 (1989).
Tomalia, D.A.: Genealogically Directed Synthesis: Starbust/Cascade Dendrimers and Hyperbranched Structures. *165*, (1993).
Tourillon, G., see Fontaine, A.: *151*, 179-203 (1989).

Ugi, I., see Bley, K.: *166*, 199-233 (1993).

Vinod, T. K., Hart, H.: Cuppedo- and Cappedophanes. *172*, 119-178 (1994).
Vögtle, F., see Dohm, J.: *161*, 69-106 (1991).
Vögtle, F., see Knops, P.: *161*, 1-36 (1991).
Vögtle, F., see Ostrowicky, A.: *161*, 37-68 (1991).
Vögtle, F., see Schulz, J.: *172*, 41-86 (1994).
Vögtle, F., see Schröder, A.: *172*, 179-201 (1994).
Vogler, A., Kunkeley, H.: Photochemistry of Transition Metal Complexes Induced by Outer-Sphere Charge Transfer Excitation. *158*, 1-30 (1990).
Vondenhof, M., see Mattay, J.: *159*, 219-255 (1991).

Wan, P., see Krogh, E.: *156*, 93-116 (1990).
Warwel, S., Sojka, M., and Rüsch, M.: Synthesis of Dicarboxylic Acids by Transition-Metal Catalyzed Oxidative Cleavage of Terminal-Unsaturated Fatty Acids. *164*, 79-98 (1993).
Wexler, D., Zink, J. I., and Reber, C.: Spectroscopic Manifestations of Potential Surface Coupling Along Normal Coordinates in Transition Metal Complexes. *171*, 173-204 (1994).
Willner, I., and Willner, B.: Artificial Photosynthetic Model Systems Using Light-Induced Electron Transfer Reactions in Catalytic and Biocatalytic Assemblies. *159*, 153-218 (1991).

Yoshida, J.: Electrochemical Reactions of Organosilicon Compounds. *170*, 39-82 (1994).
Yoshihara, K.: Chemical Nuclear Probes Using Photon Intensity Ratios. *157*, 1-34 (1990).

Zamaraev, K.I., see Lymar, S.V.: *159*, 1-66 (1991).
Zamaraev, K.I., Kairutdinov, R.F.: Photoinduced Electron Tunneling Reactions in Chemistry and Biology. *163*, 1-94 (1992).
Zander, M.: Molecular Topology and Chemical Reactivity of Polynuclear Benzenoid Hydrocarbons. *153*, 101-122 (1990).
Zhang, F.J., Guo, X.F., and Chen, R.S.: The Existence of Kekulé Structures in a Benzenoid System. *153*, 181-194 (1990).
Zimmermann, S.C.: Rigid Molecular Tweezers as Hosts for the Complexation of Neutral Guests. *165*, 71-102 (1993).